**FRANCIS CLOSE HALL
LEARNING CENTRE**

Swindon Road Cheltenham
Gloucestershire GL50 4AZ
Telephone: 01242 714600

UNIVERSITY OF
GLOUCESTERSHIRE
at Cheltenham and Gloucester

NORMAL LOAN

First published 1996

Wild Goose Publications
Unit 15, Six Harmony Row, Glasgow G51 3BA

Wild Goose Publications is the publishing division
of the Iona Community.
Scottish Charity No. SC003794. Limited Company Reg. No. SCO96243.

ISBN 0947988 80 7

Cover illustration: detail from 'Village School: Stonesfield,
Oxfordshire' © Sasha Ward. Commissioned by Oxfordshire
County Council.

Distributed in Australia and New Zealand by Willow
Connection Pty Ltd, Unit 7A, 3-9 Kenneth Road, Manly Vale
NSW 2093, Australia.
Permission to reproduce any part of this work in Australia or
New Zealand should be sought from Willow Connection.

A catalogue record for this book is available from the British
Library.

Printed by the Cromwell Press, Melksham, Wiltshire.

To my wife Anne
who has supported me throughout my career.

Foreword

Four years ago, in June 1992, the largest international conference ever held, the United Nations Conference on Environment and Development – often called the Earth Summit – was staged in Rio de Janeiro. Well over one hundred heads of government attended. Green movements, pressure groups and the media were there in strength and there was much publicity for the environment.

Two international Conventions were signed in Rio – a Framework Convention on Climate Change and a Convention on Biological diversity. Both represented large steps forward in environmental commitment on the part of the nations of the world. Both laid down important principles as a prelude to action.

Since Rio there has been much *talk* about action but not much agreement. While environmental movements have kept up the pressure, vested interests representing those who feel they might lose out have also taken the stage, arguing that preserving the environment as envisaged by the Rio Conventions is too expensive for the world to contemplate. Although the discussion has largely centred around what measures are necessary to preserve the environment and how they might be achieved, the debate has also exposed a wide variety of attitudes towards the environment and its conservation. Why indeed should we preserve the Earth and its living systems of plants, animals and other life forms?

The Green movements give various answers to these questions, sometimes supporting them by reference to religious attitudes, especially ones culled from Eastern religions. But where are the Christians in the debate? The church has certainly not been noticeable for its pronouncements on the subject.

Although many Christians broadly go along with environmental concern, few have given serious thought to the issues involved. Some, particularly amongst the evangelicals, deliberately ignore the environment since they believe it is only of subsidiary importance in relation to more spiritual matters. Others go further and unthinkingly exploit the environment and the Earth's resources, arguing their right to do so on the basis of the 'dominion' over the Earth given to our first parents in the first chapter of Genesis.

However, as these chapters point out, such attitudes are far removed from the message of the Bible regarding the relationship between God, people and the rest of creation. The Christian religion is a very earthy religion. Right at its heart is 'the Word made flesh' who came not only to save us from sin but also to free the whole of creation from the bondage which sin has brought. When, as Christians, we say 'Jesus is Lord' we recognize (although we don't often realize it) that Jesus is Lord also of the whole of creation – something that has very large implications for us.

From a lifetime's experience of working with nature in different parts of the world and with peoples who live closely with nature, Ghillean Prance eloquently introduces us to something of the wonder and excitement of nature. He also quite simply presents the Christian imperative to take responsibility for looking after God's Earth in God's way. If you are prepared to be both excited and challenged, read on.

Sir John Houghton FRS
Chairman, Royal Commission on Environmental Pollution

Contents

9   Introduction

11   The integrity of creation

27   The degradation of creation

54   From sin to stewardship

68   Is there any hope?

80   Actions for an earthkeeping church

84   Useful resources

86   Further reading

Introduction

This book presents the contents of the 1995 London Templeton Lectures which were delivered at the Linnean Society of London in January 1995.

I am most grateful to Professor Roland Dobbs for making all the arrangements for the lecture series, and also to the four people who chaired the lectures who were, in order of chapters one to four of this book, Sir Maurice Lang, the Rt Rev. Hugh Montefiore, Sir Michael Colman and Sir John Houghton.

I have included a list of some further reading on the topic of the environment and faith and also some suggestions for action by churches and communities willing to make a commitment to the stewardship of creation.

Chapter one

The integrity of creation

Four things on earth are small,
yet they are exceedingly wise:
the ants are a people without strength,
yet they provide their food in the summer;
the badgers are a people without power,
yet they make their homes in the rocks;
the locusts have no king,
yet all of them march in rank;
the lizard can be grasped in the hand,
yet it is found in kings' palaces.

(Proverbs 30:24-28)

The purpose of this chapter is to show the resplendence and the excitement of nature, God's creation which I have studied, admired and marvelled at since I was a small child growing up in the Isle of Skye in the western Highlands of Scotland. Although my family moved back to England when I was only seven years old, it was in Skye that I began my lifelong passion for natural history. Little did I know then where it was going to lead me.

In 1960 I had the privilege of taking part in an Oxford University expedition to Turkey. One of the places which we visited to collect plants was the Ak Daglari (White Mountains) of the Lycean Taurus and it was there that I had my first mountain-top experience which seems a fitting place to begin this book which considers various aspects of creation and the way in which humankind is treating it today.

The walk to the top of the Ak Daglari in the summer heat of August was quite strenuous, especially when one is carrying a rucksack of botanical collecting equipment. On the way up, as we passed through a beautiful remnant of a majestic cedar forest (*Cedrus libani*), I realized that so little of these forests remains because they have been under pressure for so long. Even the building of Solomon's temple must have caused considerable deforestation.[1] But it was good to walk through this forest, treading on the soft carpet of needles and breathing in the healthy smell of a conifer forest. Above the tree line of the mountain lies a barren scree of white limestone rock. It was amongst these rocks that we were searching for botanical treasures such as the spiny, cushion-like species of *Astragalus* and *Acantholimon*. We slowly collected our way to the top of the mountain, taking more interest in the plants than the view. Eventually we reached the summit and our reward was what is still one of the most beautiful landscapes I have ever seen in all my travels to many places. To my surprise I broke out in praise of God the creator in a strange language, my first experience of speaking in tongues. Although I had been studying plants and animals since early childhood, the beauty of this part of God's creation touched me in a special way and showed me a little bit more of the wonders of the delicately balanced planet on which we live. I could now understand a little better how 'God saw everything that he had made, and indeed, it was very good' *(Genesis 1:31)*.

This sort of experience led me to an understanding and greater appreciation of one of the great hymns of creation, Psalm 8.

O Lord, our Sovereign,
how majestic is your name in all the earth!
You have set your glory above the heavens.
Out of the mouths of babes and infants
you have founded a bulwark because of your foes,
to silence the enemy and the avenger.
When I look at your heavens, the work of your fingers,
the moon and the stars that you have established:
what are human beings that you are mindful of them,
mortals that you care for them?
Yet you have made them a little lower than God,
and crowned them with glory and honour.
You have given them dominion over the works of your hands;
You have put all things under their feet,
all sheep and oxen,
and also the beasts of the field,
the birds of the air, and the fish of the sea,
whatever passes along the paths of the seas.
O Lord, our Sovereign,
how majestic is your name in all the earth!

This psalm of praise not only reveals God to us through creation, it also shows clearly people's role in it. The dominion which is given in the first chapter of Genesis is repeated here, and so there is an awareness of the caretaking role for people in the creation which has moved the psalmist to such praise. Dominion and abuse of it by humankind will be discussed later, but before looking at the destruction of creation in the next chapter it is important to look at a few of the wonders of the natural environment and its mechanisms so that we have an understanding of the delicate balance of nature, and of its creator. Since most of my experience of nature has been in the rainforest of Amazonia rather than the mountains of Turkey, I will mainly use examples from the Amazonian rainforest where I have spent more than eleven years during the last thirty years.

To enter the majestic Amazonian forest for the first time is an awe-inspiring experience. The tall trees, many with buttressed trunks, give the forest a cathedral-like appearance. They form a dense overhead canopy which prevents all but three per cent of the sunlight from reaching the forest floor. In this permanent twilight the strangely shaped trunks, no two alike, and the bizarrely contorted stems of the vines begin to play on one's imagination. This is very different from a Middle Eastern garden of Eden, but to the naturalist, this is a paradise because of the great variety of plants and animals which the forest contains, and the fascinating way in which the different organisms interact. A few examples will demonstrate this web of life which binds together every species into an interdependent ecosystem.

The royal water lily

The royal water lily (*Victoria amazonica*) is one of the best-known plants of Amazonia. It has magnificent floating pads, up to eight feet in diameter, which can support the weight of a small child. The pads have turned-up edges which reveal spines that protect the pads from aquatic predators. The lily produces a large white flower which opens at dusk. It is an unforgettable sight to watch the mass of white stars open as darkness falls across an Amazonian lake. When the flowers open, the insides of the white flowers are up to eleven degrees centigrade warmer than air temperature, and they exude a strong fruity smell.

As soon as the flowers begin to open, large, brown scarab beetles (*Cyclocephala castanea* and *Cyclocephala hardyi*) start to arrive. They enter the central cavities of the flowers and begin to feed on starch-rich appendages called paracarpels which are at the top of the cavity. Later in the night the flowers close up tightly and trap the beetles inside. The beetles are content with food and a warm resting place, and they do not damage the vital parts of the flower because they have the food provided by the paracarpels. The flower remains closed during the next day and opens again at night. The flower looks different now, for it has changed colour completely, and when it reopens on the second night, it is dark purple and red. It is no longer scented or heated above air temperature. As the flower opens for the second time, it releases its pollen so that it adheres to the beetles as they emerge, sticky with plant juices from inside the cavity. The beetles then fly off into the night to find a white flower to enter. The stigma of the white flower is receptive and ready to receive pollen, and in this way the beetle transports pollen from plant to plant. Because a plant produces a new flower only every second day, the beetles ensure that the pollen is carried to a different plant, so there is cross-pollination.[2] What a wonderful relationship between beetles and water lilies.

The author showing the spiny and ribbed under-surface of the royal water lily.
Photo: Ghillean Prance

The Brazil nut

The Brazil nut *(Bertholletia excelsa)* has a complicated flower structure. The stamens which form a ring at the centre of the flower are covered by a tightly closed protective hood over the top of the flower. Nectar is produced inside the hood at the base of a large number of sterile appendages called staminodes. The hood, which protects the flower parts from ready access by small insects, can be lifted only by a large, strong insect which can push up against the springlike action of the ligule which attaches the hood to the basal ring of stamens. Only large bees, such as the carpenter bee *(Xylocopa)*, large bumble bees *(Bombus sp.)*, and the orchid bees *(Eulaema)* and the other large *Euglossinae*, can lift up the hood.[3] The bee lands on the top of the hood and finds its way into the crack to forage for nectar inside. At the same time the pressure of the spring forces the back of the bee hard against the pollen-producing stamens and the stigmatic surface in the centre of the flower. The bee is then dusted with pollen and will transport it to another flower as it continues its foraging route.

Brazil nuts are produced in a large, round, hard, woody fruit about twice the size of a cricket ball. The fruits, which take fourteen months to develop after pollination, fall to the ground in January and February during the rainy season. The fruit case is so hard that it can remain for several years without rotting, whereas the seeds or nuts, arranged inside like the segments of an orange, would lose their viability if not exposed. However, nature has arranged a mechanism to open these hard fruits. A large rodent, the agouti *(Dasyprocta sp.)*, chews open the outer shell of the fruit and removes the nuts. It then buries caches of Brazil nuts away from the tree. Because the agouti does not find all its caches, the nuts are dispersed around the forest. The production of one of the most valuable products of the Amazon forest is therefore dependent on large bees which visit the top of the tree above the canopy and on a rodent which forages on the forest floor.

The calabash

The calabash *(Crescentia cujete)* is the most useful utensil of the Amazon region. The Indians and the local peoples use this fruit as a water pot, soup bowl, canoe bailer and even as a hat. The mature fruit can reach the size of a football and has an extremely hard outer shell which is why it is so useful. The green flowers of the calabash are produced on the trunk and branches of the tree and are therefore easily accessible to the bats which visit the flowers to sip their nectar and in the process cause pollination.

The flowers are not, however, the only nectar producing organs of the calabash. The young developing fruit is soon covered by minute whitish dots which exude abundant nectar. As soon as this process starts the fruit is attended by numerous aggressive ants which rigorously defend their source of food from any other marauding animal. As the fruit goes through its development it is protected by ants and as it matures the nectaries dry up and the fruit becomes hard and woody. It is now protected mechanically and no longer needs the ants and it is this hard woody shell which makes it such a useful object.[4] This multipurpose fruit prospers along Amazonian riversides because of the bats which pollinate it and the ants which protect it through adolescence. I can't help thinking of the book of Proverbs 6:6:

Go to the ant, you lazybones; consider its ways and be wise.

The calabash (Crescentia cujete)
Photo: Ghillean Prance

Costus

The genus *Costus*, which is related to the ginger plant, is often cultivated as an ornamental in tropical gardens. This is because of its mass of brightly coloured specialized flower–leaves (bracts). These bracts also secrete nectar and are tended by protective ants. If *Costus* is grown where ants are excluded the seed set drops by over eighty per cent because another insect lays its eggs in the ovules of the flowers. The ants drive this insect away but they do not interfere with the larger hummingbirds which visit the flowers and insert their beaks into the flower tube to drink the floral nectar and at the same time become dusted with *Costus* pollen. This beautiful ornamental plant depends on hummingbirds and ants for its existence.

The leaf-cutter ant

A common sight on the floor of the Amazon forest is a trail of leaf fragments moving along the ground. Closer examination shows that the pieces of leaves are moving because they are carried by ants which have snipped them off nearby trees. The ants transport the leaves to gigantic, complex underground chambers which they inoculate with a certain species of fungus. The growing mass of fungal mycelia provides food for the ants, thus making them miniature underground gardeners. The fungus is just a mass of unidentifiable, threadlike strands.

In observing these ants I found that if a random selection of forest leaves were placed on the trail, the ants would chew up some of these leaves and take them to their nests, whereas they would simply clear others off the trail and reject them as trash. Some of the rejected leaves have been found to contain natural fungicides, and thus one of the best ways of discovering new fungicides is to observe the behaviour of the leaf-cutter ant. The delicate chemical receptors of the ants, which can detect substances that would destroy their food garden, give clues about which plants contain these natural fungicides.

The cacique bird

Many travellers have observed the curiosity of a wasp or hornet nest close to a colony of cacique birds (*Cacicus*), which weave long, pendulous, basketlike nests. The birds choose to build near the hornets because these insects drive off botflies whose larvae can kill the cacique chicks. Nature has developed a protective relationship for the chicks with an insect which we, from a human point of view, commonly regard as noxious.

The story is not as simple as that, however, because the giant cowbird (*Tordo pirata*) lays its eggs in cacique nests, in much the same way as the cuckoo in Europe.

Neal Smith, a zoologist from the Smithsonian Tropical Research Institute in Panama, noticed that some colonies of caciques reject the interloping cowbird's egg, but others allow the parasite chick to hatch and grow. Smith eventually found that caciques nesting with the hornets throw the cowbird eggs out, whereas those without the insect protectors rear the cowbird chicks.

He then found that cacique chicks in a nest with a cowbird chick are nine times less likely to have botflies than those growing up without cowbirds or hornets. The reason is that cowbird chicks soon open their eyes and are much more aggressive than the cacique chicks. They defend themselves and the cacique chicks from botflies, and we thus have a complex pattern of behaviour involving two species of bird and several species of hornets and wasps.

The three-toed sloth

The three-toed sloth *(Bradypus tridactylus)* is not just a slow-moving, lazy mammal; it is a mobile, multi-organism ecosystem. It is well camouflaged because of the green algae which live on its grey fur, and this helps to protect it from its main enemy the harpy eagle *(Harpia harpyja)*. However, the sloth is also host to beetles, ticks, and mites which live protected in its fur. The most interesting hitchhikers are a species of moth. The sloth descends from the trees to defecate only occasionally (once in three weeks). One species of moth lays its eggs in the sloth dung where the larvae develop until the mature moths link up to their home in the fur of the sloth on its next ground visit.

Peoples of the Amazon

For at least the last ten thousand years a large number of indigenous peoples, the Amazonian Indians, have inhabited the Amazon region. By the time Europeans reached the Amazon in the early sixteenth century, there was a population of at least five million Indians speaking as many as three hundred different dialects derived from four major stem languages – Tupi, Arawak, Carib and Pano-Tucanon.

Today the remnant Indian population is only about a quarter of a million since their territories have gradually been occupied by settlers, miners and loggers.[5]

At the time of the arrival of white settlers, the Amazon region sustained a large population and was largely a forest with an abundance of species. In other words this population had minimal impact on the environment. Early explorers recounted the abundance of manatees, turtles and caiman. Their description of the quantity of these animals along the rivers is quite unlike that of today, which is that the animals are rare and difficult to see.

The three-toed sloth (Bradypus tridactylus)
Photo: Ghillean Prance

Today one could not echo the words of Henry Bates describing the lake margins of the Rio Japurá region in 1850: 'This inhospitable tract of country ... contains in its midst an endless number of pools and lakes tenanted by multitudes of turtles, fishes, alligators and water serpents'.[6] Yet we know that the Indians hunted these species as sources of food, oil and many other products. Today, the manatee or sea cow *(Trichechus unguis)* is one of the rarest and most endangered of all Amazonian species.

Remnants of charcoal in many Amazonian soils show that the Indians did cut and burn forest for their plantations of cassava, corn and other crops. However, they cut small patches which were surrounded by forest so that when a field was abandoned, the seeds were available for the forest to regrow, and the soil microorganisms were also nearby to recolonize the soil. In other words, the Amazon Indian agriculture, which developed gradually over several millennia, preserved the soils, the wildlife and the ecosystem as a whole, but at the same time they were able to enjoy its cornucopia of fruits, game, building materials, medicines and many other forest products.

Yanomami Indians in the area they have cleared to plant their crops. The small areas which they cut and burn soon regenerate into forest, doing no permanent damage.
Photo: Ghillean Prance

A view of deforestation along the Cuiabá-Santarém highway.
A view typical of the 1980s in the Amazon, when vast areas of
rainforest were felled to create cattle pasture.
Photo: Ghillean Prance

Today scientists are studying the ecology of the Amazonian Indians so that we might better understand how they have managed not only to survive but also to prosper in the Amazon forest. The Kayapó Indians of the village of Gorotire practise the traditional swidden agriculture where they cut and burn small discrete areas of forest to grow their crops. However, they also manage areas which appear to the uninitiated as virgin forest.[7] For example, they plant fruit trees along the trails which they use most frequently. These plants are therefore accessible as food on long journeys and are easy to manage as they travel.

The Bora Indians of Amazonian Peru make extensive use of their 'abandoned' fields.[8] The fields for their primary crop were thought to be abandoned after a few years of use. Research has shown, however, that Indians often control the succession of regeneration so that a large quantity of useful plants are encouraged to recolonize the old fields. Over the years the Bora make frequent visits to these old fields to harvest fruits, medicinal plants, edible mushrooms, fibres, fish poisons and many other products.

The indigenous peoples are an integral part of the Amazonian Eden, a part that has much to teach people from modern industrialized society. Water lilies, Brazil nuts, calabash fruit, beetles, bees, ants, hummingbirds, caciques, hornets, sloths, moths and a host of other organisms make up the connected web of biodiversity that is the Amazonian rainforest.

I could extend this catalogue of the marvels of creation around the world from the depths of the oceans to the tops of mountains, and certainly to the Scottish Highlands. Here we have had a glimpse of what must have had a great artisan behind it all. It is hardly surprising that when God had created the heavens and the earth he reviewed it and concluded that it was good. 'God saw everything that he had made, and indeed, it was very good' (*Genesis 1:31*). And the Apostle Paul saw God revealed in creation as well as through his other experiences such as the Damascus road. 'Ever since the creation of the world his external power and divine nature, invisible though they are, have been understood and seen through the things he has made. So they are without excuse.' (*Romans 1:20*)

This wonder of creation and experience of the creator is what I first saw as a young boy roaming the moors of Scotland, and later on a mountain top in Turkey, amongst the icebergs of Antartica and throughout the Amazonian rainforest. In Genesis 2:9 it says, 'Out of the ground the Lord God made to grow every tree that is pleasant to sight and good for food.'

Here the aesthetic component of nature is clearly given value in that the beauty of the tree is to be appreciated. Creation was intended for both aesthetic and utilitarian purposes, it was not all destined for destructive use and abuse. People were put into it to look after it rather than destroy it.

One of my favourite passages of scripture is the last few chapters (38-41) of the book of Job. After Job has been through extreme hardship and suffering and has been counselled by his friends who were false comforters, God finally reveals himself out of a whirlwind to him. Instead of making a deep theological pronouncement, God asks Job a question: 'Where were you when I laid the foundation of the earth?' (*Job 38:4*). He proceeds to point out Job's ignorance of creation and then gives him a detailed description of some of the marvels of nature of the Middle East and of the Universe. God asks Job:

Who determined its measurements – surely you know!
Or who stretched the line up on it?
On what were its bases sunk,

or who laid the cornerstone,
when the morning stars sang together
and all the heavenly beings shouted for joy?
Or who shut in the sea with doors
when it burst out from the womb?
when I made the clouds its garment,
and thick darkness its swaddling band,
and prescribed bounds for it,
and set bars and doors,
and said, 'Thus far shall you come, and no farther,
and here shall your proud waves be stopped'?

(Job 38:5-11)

This passage goes on to describe many of the physical and natural features of creation. We read of light, darkness, rain, snow, hail, ice, clouds, lightning and even about the constellations Pleiades, Orion and Ursa major. This description of the physical is followed by a wonderful lesson about animal behaviour where we read about how the lioness tends her cubs, the mountain goats give birth, the calving of deer, the way in which the wild donkey ekes out an existence foraging in barren lands, the strength of the wild ox, the habit of the hen ostrich abandoning her eggs, the might and majesty of the horse, the soaring hawk and the nesting of the eagle. God's next question to Job is:

Is it at your command that the eagle mounts up
and makes its nest on high?
It lives on the rock and makes its home
in the fastness of the rocky crag.
From here it spies the prey;
its eyes see it from far away.
Its young ones suck up blood;
and where the slain are, there it is.

(Job 39:27-30)

As God calls Job to pull himself together, the lesson in natural history continues with a description of two of the giants of creation – Behemoth the hippopotamus and the strength he gets from eating grass, and the mighty Leviathan, the whale.

Look at Behemoth, which I made just as I made you;
it eats grass like an ox.
Its strength is in its loins,
and its power in the muscles of its belly.
It makes its tail stiff like a cedar;
the sinews of its thighs are knit together.
Its bones are tubes of bronze,
its limbs like bars of iron.
It is the first of the great acts of God –
only its Maker can approach it with the sword.
For the mountains yield food for it
where all the wild animals play.
Under the lotus plants it lies,
in the covert of the reeds and in the marsh.
The lotus trees cover it for shade;
the willows of the wadi surround it.
Even if the river is turbulent, it is not frightened;
it is confident, though Jordan rushes against its mouth.
Can one take it with hooks,
or pierce its nose with a snare?

(Job 40:15-24)

The lesson from these chapters of Job is that true repentance in-volves recognition that all we have is lent to us in trust and that all created things are God's, entrusted to our care even though we may use them.

As a biologist I have concentrated in this chapter on the living, natural side of creation, but it is interesting and important that the passages from Job to which I have referred are divided almost equally between the wonders of nature and of the physical world. As we look at the destruction of creation in the next chapter I will concentrate on the physical features which hold our world to-gether. A study of the atmosphere, of oxygen, the carbon cycle and of the way in which the toxic gas ozone protects us from excessive ultraviolet radiation can equally well demonstrate the wonders and the delicate balance of creation. The flow and use of energy from the sun and the way in which the correct level of carbon dioxide in the atmosphere keep our planet at a tempera-ture which sustains life is of equal wonder. The soil which sus-tains most terrestrial plant life is an extraordinary mixture of the physical and the living. The life-sustaining soil has been built up over millions of years and is a most precious part of creation which we are fast eroding away. The cycle of nutrients through the soil, the interaction of thousands of species of microorganisms with the dead leaves and with the minerals in the soil epitomize the delicate balance of nature which maintains life on earth. Creation is full of wonder and integrity. Isaiah the prophet said:

> *I will put in the wilderness the cedar,*
> *the acacia, the myrtle, and the olive;*
> *I will set in the desert the cypress,*
> *the plane and the pine tree together,*
> *so that all may see and know,*
> *all may consider and understand*
> *that the hand of the Lord has done this,*
> *the Holy One of Israel has created it.*

(Isaiah 41:19-20)

May we see, know and understand the wonders of God's creation, and from that understanding, from both our science and our theology, act upon it to be better stewards of creation.

Notes

1 The building of the temple as described in 1 Kings chapters 5, 6 and 7 refers to the large quantities of cedar and cypress wood being used. For example:

'Therefore command that the cedars from the Lebanon be cut for me. My servants will join your servants, and I will give you whatever wages you set for your servants; for you know that there is no one among us who knows how to cut timber like the Sidonians.' *(1 Kings 5:6)*

'Hiram sent word to Solomon, "I have heard the message that you sent to me; I will fulfil all your needs in the matter of timber and cypress."' *(1 Kings 5:6)*

'He lined the walls of the house on the inside with boards of cedar, from the floor of the house to the rafters of the ceiling, he covered them on the inside with wood; and he covered the floor of the house with boards of cypress.' *(1 Kings 5:8)*

'It was roofed with cedar on the forty five rafters, fifteen in each row, which were on pillars.' *(1 Kings 7:3)*

2   Ghillean Prance and Jorge Arias, 'A study of the floral biology of *Victoria Amazonica* (Poepp.) Sowerby (Nymphaeaceae)', *Acta Amazonica*, no. 5 (1975) pp.109-39.

Ghillean and Anne Prance, 'The beetle and the water lily', *Garden Journal*, no. 26 (1976) pp.118-21.

3   Bruce Nelson, Maria Absy, Eduardo Barbosa, Ghillean Prance, 'Observations on flower visitors to *Bertholletia excelsa* H. & B. and *Couratari tenuicarpa* A.C.Sm. (Lecythidaceae)', *Acta Amazonica* Supl. 15 (1-2) (1985) pp.225-34.

Ghillean Prance, 'The pollination and androphore structure of some Amazonian Lecythidaceae', *Biotropica*, no. 8 (1976) pp.235-41.

4   Thomas Elias and Ghillean Prance, 'Nectaries on the fruit of *Crescentia* and other Bignoniaceae', *Brittonia*, no. 30 (1978) pp.175-81.

5   William Denevan (ed.), *The Native Population of the Americas in 1492* (Madison: University of Wisconsin Press, 1976).

6   Henry Bates, *The Naturalist on the River Amazons* (Reprinted London: J.M. Dent & Sons, Aldine Press, 1969), p.315.

7   Darrell Posey, 'The Keepers of the Forest', *Garden*, no. 6 (1982) pp.18-24.

Darrell Posey, 'A preliminary report on diversified management of tropical forest by the Kayapó Indians of the Brazilian Amazon', *Advances in Economic Botany*, no. 1 (1984), pp.112-26.

8   William Denevan and Christine Padoch (eds.), 'Indigenous agroforestry in the Peruvian Amazon', *Advances in Economic Botany*, no. 5 (1988) pp.1-107.

Chapter two

The degradation of creation

Each person on Earth should have free access to air, water, education and contraception.

(Olivia Gherman, addressing the World Population summit in Cairo, 1994.)

Since all have sinned and fall short of the glory of God.

(Romans 3:23)

The environmental crisis of today is so great that even scientists are acknowledging that they can't solve the problems alone. For example, at a recent 1993 population summit of fifty-eight of the world's scientific academies, including the Royal Society and the US National Academy of Sciences, it was acknowledged that science alone could not solve the population crisis: 'Furthermore it is not prudent to rely on science and technology alone to solve problems created by rapid population growth, wasteful resource consumption and poverty. Scientists, engineers, health professionals should study and provide advice on: cultural, social, economic, religious, educational and political factors that affect reproductive behaviour, family size and successful family planning.'[1]

A revision of the world conservation strategy of the International Union for the Conservation of Nature and Natural Resources, the Worldwide Fund for Nature and the United Nations Environmental Programme entitled 'Caring for the Earth' clearly stated the need for a new ethic for sustainable living and conservation of biological resources: 'Establishment of the ethic needs the support of the world's religions because they have spoken for centuries about the individual's duty of care for fellow humans and of reverence for divine creation'.[2]

In a recent book, ecologist Lawrence Hamilton said, 'It is not ecologists, engineers, economists or earth scientists who will save spaceship earth, but the poets, priests, artists and philosophers'.[3] The organizer of the 1992 Earth Summit in Rio de Janeiro, Brazil, Dr Maurice Strong, gave a lecture at the Royal Botanic Gardens,

Kew in 1993 in which he concluded: 'In the final analysis, our economic and social behaviour is rooted in our deepest moral and spiritual motivations. We cannot expect to make the fundamental changes needed in our economic life unless they are based on the highest and best of our moral, spiritual and ethical traditions, a reverence for life, a respect for each other, and a commitment to responsible stewardship of the Earth. The transition to a sustainable society must be undergirded by a moral, ethical and spiritual revolution which places these values at the centre of our individual and societal lives.'

It is in this broad context that I present here a brief outline of five of the most serious aspects of the environmental crisis: population growth, global warming, pollution, the loss of biodiversity and the loss of soil. I hope that, rather than cause alarm, it will provoke Christian thought and action based on a strong theology of creation and of justice.

Population

I am focusing on the subject of population first because it is the fundamental root of the environmental crisis and the most important issue to address if there is to be any future for humankind. The issue of population is also the one which the Church tends to ignore most or even oppose discussing because of the ethical issues involved. The basic statistics about population growth are simple and obvious, but the consequences of ignoring them are dire. It took two million years from the origin of the human species for the worldwide population to reach one billion in about 1830, just a few years after Thomas Malthus published his prophetic work 'An Essay on the Principle of Population'. Over the next century the figure doubled to two billion in 1930, but by then growth had become exponential, growing very rapidly because each new birth multiplied the problem. Our population reached four billion in 1970 and five billion in 1987. The overall growth rate has fallen in recent years from 2.1 per cent to 1.7 per cent, but this is still exponential and we can predict double that amount in forty years' time. Therefore, we can expect a population of around ten billion in the year 2030 unless drastic measures are taken to slow down the growth rate. Between seven and eight per cent of all human beings that ever lived on this planet are alive today! In many non-industrialized countries, fifty per cent of the population are under reproductive age, which will lead to a future burst in population growth. In Africa, sixty per cent are younger than twenty years old and half are unemployed, with no future job prospects. There is considerable disagreement about the future growth of human population and about when and what will stop it growing, but it is generally predicted to stabilize soon after the year 2100.

Apart from the sociological, political, economic and environmental consequences of excess population that I will discuss later, humankind is ignoring the basic biological fact that any organism whose population rises above its resource capacity is bound to crash. Many studies of population biology, whether they are with single-celled paramecium, fruit flies, rats or deer, have repeatedly shown that species whose population increase with excessive rapidity are also subject to rapid declines or crashes in their numbers.[4] The law of biology is that there are limits to growth. We know that the human species is subject to the same constraints and biological rules as these other species. Our intellect and our power of reasoning that invented agriculture and the industrial revolution, the main causes of population growth, should also enable us to control our destiny and avoid a population crash. But are we too arrogant? The greatest challenge of the next forty years is the stabilization of world population. This is often an unpopular subject with governments and religions, but it is vital that it is not ignored.

Increased population leads to the shortage of resources and hence to social unrest and fighting over their distribution, to destruction of the natural environment through the greater demand for food and other products, to greater use of energy and consequently greater pollution, to increase in carbon dioxide and loss of the ozone layer which protects us from excess ultraviolet radiation, and eventually to the extinction of many biological species.

I have spent considerable time amongst the Yanomami Indians and have personally observed the truth of Napoleon Chagnon's 1968 study[5] which showed that when a village is small, with a population of around eighty people, it is quite peaceful. By the time there are 120 people, there are many disputes and fights; and when the population reaches 180 to 200, there is usually a huge fight that ends up with the village splitting into two. In the vast Amazon rain forest there is still room for this group fission and population increase, but this is a microcosm of what could happen today as the world's population and individual countries expand beyond their resources. For many generations we have just occupied more space or, as in America, moved west. Today there is no longer anywhere to go on Earth.

Already life in many of the world's large cities is deteriorating badly because they are just too large to manage ecologically. São Paulo in Brazil had 20,000 inhabitants in 1818 and currently has seventeen million. The nineteen million inhabitants in Mexico City now live in one of the most polluted atmospheres in the world and misery is on the increase in Calcutta, Karachi, Lima, Manila and many other large cities. El Salvador is the most overpopulated and environmentally degraded country in Latin America. It is not surprising that this country has been in a continual state of strife since 1969. Rwanda is Africa's most densely populated country and it is also in a state of civil war. Despite the current signs of the end of the period of Cold War, local and regional warfare and environmental deterioration will continue if we have not resolved the issue of population growth within the next twenty years. Is it beyond hope that we could achieve a stable population of nine billion by the middle of the twenty-first century? To achieve this there would need to be an average fertility rate of 1.7 children per woman from early next century.

The 1993 population conference of the world's scientific academies concluded that global social, economic and environmental problems cannot be solved without a stable world population and set a goal to reach zero population growth within the lifetime of their children. Their concluding statement was: 'Let 1994 be remembered as the year when the people of the world decided to act together for the benefit of future generations'.[6]

A recent report from the American Association for the Advancement of Science states that 'To do nothing to control population numbers is to condemn future humans to a lifetime of absolute poverty, suffering, starvation, disease and associated violent conflicts'.[7] Can we rise to this challenge? Can the church also respond to this?

In September 1994 the United Nations held a population conference in Cairo. In spite of opposition from the Vatican and some fundamentalist Moslem countries, the conference made great progress and allocated $17 billion for future programmes. The Conference was also notable for the active participation of the United States after the opposition to population control of the Reagan and Bush governments. One of the main conclusions of the conference was that availability of education and freedom of choice for women are crucial factors in decreasing family size.

It is a proven fact that access to education and employment for women, and increased life expectancy so that more children are not needed as a means of social security, are all factors that lead to a reduction in population growth. It is estimated that about 100 million couples who would like to use contraceptives are denied access to them. These are the areas upon which we should concentrate over the next few decades. The problems of overcrowding can also be considerably reduced if there is a much more equitable distribution of wealth and if we can address the issues of social justice. But where is the church in these matters?

Ninety million people are added to the world each year which is equivalent to adding another Mexico or three Canadas to the world annually. Can the world sustain such growth? Where are the resources to provide food, water and even contraception to them? How much additional resources will they consume, and how much extra pollution will they cause? The 1995 'State of the World' report suggests that the world will have a massive food shortage by 2030, just thirty-five years ahead.

During the last hour when you were reading this book 10,800 new mouths to feed were born. At this rate over one million are added to our population every four days yet there are already 384 million starving people in the world and another 595 million suffer serious calorific deficiencies. Can a Christian remain complacent about the issues of population? There is no room for complacency. Unless we want complete anarchy to rule the future world, we must contain population growth.

In case you think this is only a problem of the non-industrialized world we should look at the result of population growth in the UK on the environment. Our population is still increasing by 116,000 a year. Each new Briton uses at least thirty times more fossil fuel than a Bangladeshi. It would therefore take 3.39 million Bangladeshis to have the same environmental effect as our apparently low increase.

The impact of population growth is severely compounded by the uneven distribution of reserves around the world, and growth in the industrialized countries has a much greater effect on resources than that of the non-industrialized world. There *is* enough food in the world to feed the present population if it were to be evenly distributed. However, the present situation is that one sector of the population feeds lavishly and drives around in large cars, while about one billion people are eating less food than their bodies require and some 500 million of them are chronically hungry.

Economist Lester Thurow, Dean of the Sloan School of Management of the Massachusetts Institute of Technology said, 'If the world's population had the productivity of the Swiss, the consumption habits of the Chinese, the egalitarian instincts of the Swedes, and the social discipline of the Japanese, then the planet could support many times its current population without privation for anyone. On the other hand, if the world's population had the productivity of Chad, the consumption habits of the United States, the non-egalitarian instincts of India and the social discipline of Argentina then the planet could not support anywhere near its current numbers.'[8]

Climate change

In spite of conflicting reports about the extent of the greenhouse effect, mainly by politicians trying to evade the issue, its reality is beyond doubt. The increase of the so-called greenhouse gases in the atmosphere is causing a definite warming trend in the earth's atmosphere. Like the panes of glass in a greenhouse, gases such as carbon dioxide, halocarbons, methane and nitrous oxide allow most solar radiation to enter the earth's atmosphere, but prevent part of the heat being re-radiated back into space. The trapped heat causes a gradual warming in the atmosphere. Whether the climate change over the next century is 1.5°C or 4.5°C is academic; even a change of 1.5°C – the estimate of the more conservative scientists – will have serious repercussions for the future. Britain's mild November (1994) was no local phenomenon, but part of the hottest spell of worldwide weather ever recorded. Figures presented at a US government conference in Washington in November 1994 showed that the period March to October 1994 was about 0.4°C above normal, making it the hottest period since records began in the 1860s. The decade of the 1980s was 0.2°C warmer than the average for 1950-1980 and 0.5°C warmer than a century ago. The year 1990 was the warmest on record and 1991 the second warmest, while 1988 and 1987 were almost as hot. 1994 was not quite the warmest because of a very cold January and February. 1992-1993 were slightly cooler because of a lull in the warming trend caused by the massive eruption of the volcano Mount Pinatubo in the Philippines in the summer of 1991. Now that the effects of the eruption have died away we are returning to the normal warming trend. This means that 2030 is likely to be at least 0.5°C warmer than 1990. In spite of these clear warnings of the future, politicians are still complacent about addressing the issues because of the cost of correction.

Since pre-industrial times, atmospheric carbon dioxide concentration has risen from 280 parts per million by volume to 354 ppmv. The rate of emission continues to rise, principally due to the burning of fossil fuels. Half of the carbon dioxide content added to the atmosphere during the entire human history has been added during the past thirty years. This is why the greenhouse effect has happened so suddenly and requires such urgent action.

Chlorofluorocarbons are industrial chemicals which have been used mainly in refrigerators and air conditioners. These are 25,000 times more serious as a greenhouse gas than carbon dioxide, and therefore essential to control, not only because of their effect on the ozone layer but also because of their role in the greenhouse effect. Methane is thirty times more effective as a greenhouse gas, and it has risen from 800 ppbv in 1880 to 1600 ppbv today. It is therefore necessary to curb the percentage of natural gas that escapes into the atmosphere through leaks in the gas piping throughout the world, including the UK, and especially in Russia and Eastern Europe where gas pipes are notoriously leaky.

There is, however, some good news on the horizon in that the emissions of methane stabilized in 1993 and this is thought to be mainly because of repairs to leaking natural gas pipelines in the former Soviet Union. A recent report from NASA's Langly Research Center in Virginia also shows that the atmospheric content of carbon monoxide has declined slightly since 1991.[9]

If the causes of the greenhouse effect are not corrected within the next forty years, the consequences for global ecology will be serious. Not only will the climate continue to warm, but the pattern of climates in the world will alter. For example, it is predicted that the breadbasket of the United States, the Midwest, will become considerably drier and no longer suitable for such intensive agriculture. Changes in climate pattern would also cause extinction of many plant species because they would be unable to migrate at the speed of these man-made climate changes, which are far more rapid than natural climate cycles.

The warmer climate is already causing partial melting of the polar ice caps, and forcing the sea level to rise through the thermal expansion of water which takes up more space when it is warm than when it is cold. Within the next forty years, sea levels could rise by a much as 20cm, causing huge problems in low-lying areas where many of the world's major cities lie. For example, it could render the Thames Barrier ineffective, flood large areas of Bangladesh and Florida and submerge many oceanic islands.

Since warming is likely to be much greater at higher latitudes than the average increase for the whole globe, the danger of melting of the polar ice caps is very real. During the warm spell of 1994 there was a rapid spread of flowers and grasses in Antarctica. In 1995 the warming trend resulted in the break up of the Larsen ice shelf in Antarctica creating a giant new iceberg roughly the size of Oxfordshire. Another ice shelf which used to occupy Prince Gustav Channel, connecting James Ross Island to the Antarctic Peninsula, has also disintegrated for the first time in history. Yet, despite these facts, many people still deny that climate change is real.

Some action is already being taken to slow down the greenhouse effect such as more energy-efficient car motors and the introduction of solar and wind-generated energy in some places, but much more action is needed. Already a certain amount of warming is inevitable, and whatever corrective measures we take today, the warming tend will continue over the next forty years. The actions we take during this period will be crucial for the future and will determine the total extent of the greenhouse effect. This is the time in which the corrections must be made. Seventy-eight per cent of the world's energy is derived from the burning of fossil fuels, and the United States is by far the largest consumer, followed by Russia and Eastern Europe. The non-industrialized countries presently use less, but their plans for development predict a considerable increase. They regard energy-use restrictions as another way for the industrialized world to maintain its dominance over the less industrialized world, adding to the North–South conflict. However, the entire globe is in this crisis together, and global action will be needed which encourages the industrialized world to reduce consumption and the non-industrialized world to move forward in energy-efficient ways. It is unreasonable to request that the Amazonian countries halt deforestation if industrialized countries refuse to reduce their emissions of greenhouse gases.

One of the most positive outcomes of the 1992 Earth Summit held in Rio de Janeiro was the Convention on Climate Change, the aim of which was 'stabilization of greenhouse gas concentrations in the atmosphere at a level that would prevent dangerous anthropogenic interference with the climate system'. The terms of the convention were worded vaguely to avoid argument between the signatory nations, but they did commit industrialized nations to the goal of stabilizing emissions of carbon dioxide, methane and nitrous oxide at 1990 levels by the year 2000.

A computer-model analysis of this scenario by Mick Kelly and Susan Subak[10] of the Climatic Research Unit at the University of East Anglia has shown that to reduce emissions of the industrialized nations to 1990 levels would only reduce the warming trend by nine per cent. In addition the Convention does not stop the projected growth in emissions from the non-industrialized world as they too follow the trend towards industrialization. Kelly and Subak calculate that a worldwide reduction in emissions of forty per cent would be needed to control the greenhouse effect.

They correctly point out: 'It should be noted, though, that given the aspirations of the developing world this could only be achieved with massive resource transfers from North to South. The same end might be more equitably achieved through deeper cuts in northern emissions, allowing a continual rise in the southern contribution'.[11] In other words correction of the greenhouse effect involves major issues of social justice. At a time when many industrialized nations are reducing their foreign-aid programmes we should be increasing them enormously to address the issues of climate change.

Many actions are needed to slow down the greenhouse effect. In addition to population control, the following are some of the solutions we should be moving towards:

- A complete ban on the use of chlorofluorocarbons (CFCs) because of their dual role as the most powerful greenhouse gases and destroyers of the ozone layer.

- A rapid switch to greater use of energy sources which do not emit greenhouse gases, such as solar, wind, hydro and tidal power, and perhaps even a new generation of safer nuclear reactors in countries which can control the waste adequately, especially if we can harness the use of fusion rather than fission since the former would produce less waste. The use of methane for power plants emits much less carbon dioxide than coal, and methane from landfill sites could be used rather than allowing it to escape and thereby increase the greenhouse effect. So far methane has only been used for small-scale projects, particularly in India.

- The reduction of emissions of carbon dioxide through energy-conservation measures. More countries should follow the example of Japan where petrol is more expensive than elsewhere and, as a result, cars tend to be small and efficient; public transport is used more; and less carbon dioxide is sent into the atmosphere per person transported. In the future, energy produced in ways which pollute the environment must cost us more to pay for methods of cleaning up the problem.

- The halting of deforestation and the encouragement of replanting forest in both tropical and temperate regions.

- In future, the idea of a carbon tax against the producers of carbon-dioxide emissions should be taken seriously and the funds raised used to promote carbon-dioxide-free energy, particularly for the non-industrialized world, which must not be too constrained by the system. The living standards of non-industrialized countries could be raised without large increases in energy consumption, but it will cost the industrialized countries a great deal in technical aid to implement such energy-efficient economies.

– Energy conservation is essential through the use of smaller, fuel-efficient cars, well-insulated buildings, more public transport and passive solar energy in buildings.

Solar space heating collector at the Centre for Alternative Technology
Photo: CAT

Solar panel in a roof of a house in Greece used to heat water. Much more use of solar energy could be made both to heat water and to generate electricity with panels of photovoltaic cells.
Photo: Ghillean Prance

Water-powered cliff railway
at the Centre for Alternative Technology
Photo: CAT

Windmill on Orkney used to
generate electricity for the local
power supply. Wind farms are
on the increase and already
contribute to the national grid,
especially in Cornwall and
Wales.
Photo: Ghillean Prance

Solar and wind-powered telephone box
Photo: CAT

Solar and wind-powered vaccine refrigerator destined for Eritrea
Photo: CAT

I was delighted to read the conclusions of Britain's Royal Commission on Environmental Pollution and Transport which presented a report on transport in October 1994. The two-year study came down firmly on the side of more public transport and less use of private cars and road transport. One of the recommendations of the report was to cut national emissions of carbon dioxide to eighty per cent of the 1990 levels by 2020. Another recommendation was to increase the proportion of rail freight from 6.5 per cent to ten per cent by 2010. If the report is properly implemented it will not only mean that Britain is taking climate change seriously, it will also considerably improve air quality in many parts of the country because road transport is the most important source of the majority of air pollutants in our country and is the source of more than one fifth of our output of carbon dioxide. It is most unfortunate that this report has received only lukewarm response from the government.

We would do well to heed the words of US Secretary of State James Baker third, who in his maiden speech to the International Panel on Climate Change (IPCC), said, 'We face the prospect of being trapped on a boat that we have irreparably damaged – not by the cataclysm of war, but by the slow neglect of a vessel we believed to be impervious to our abuse .... The political ecology is now ripe for action.'

If we are to survive as a species, the political aspects of the greenhouse effect must be resolved during the next decade.

The ozone layer

In the lower atmosphere, ozone is known as a harmful industrial pollutant which is a component of smog that causes difficulty with breathing and reduces the growth of trees. However, in the upper atmosphere, ozone is an essential component that protects life on earth by reducing the amount of ultraviolet radiation. In 1985, scientists Joe Farman, Brian Gardiner and Jonathan Shanklin from the British Antarctic Survey discovered a large hole in the ozone layer over Antarctica. Three years later they traced the cause to chlorofluorocarbons (CFCs) which are the gases used in refrigeration, air conditioners, aerosols and various foam materials. Since this discovery, the news has become progressively worse. In October 1991, the ozone hole reached what was then a record depth extending over 21 million square kilometres, an area four times the size of the United States, and allowing twice as much ultraviolet light to reach the earth's surface. It is already seriously affecting the countries of Argentina, Chile, Australia and New Zealand. It is well known that ultraviolet radiation causes skin cancer, but now reports are beginning to come in from Chile of blind rabbits and salmon and deformed tree buds. Also in 1991 the United Nations Environmental Programme and the World Meteorological Organization announced that for the first time the ozone shield is thinning over northern temperate latitudes such as Scandinavia in summer,[12] exposing people and crops to a larger dose of ultraviolet light just when they are most vulnerable because of increased hours of sunlight and less cloud cover.

In 1992, ozone loss in the northern temperate regions was twice as much as expected. The ozone levels over northern Europe were more than ten per cent below the long-term average values in the spring of 1994, and in the winter 1993-94 measurements over the South Pole showed the lowest concentration of Antarctic ozone ever recorded.[13]

Recent figures from the US National Oceanic and Atmospheric Administration showed that on 12 October 1993 the atmosphere between the altitudes of fourteen and nineteen kilometres was completely devoid of ozone. This was caused both by the continued release of CFCs combined with the volcanic eruption of Mount Pinatubo in the Philippines. Interestingly, while this eruption slowed down the greenhouse effect for a few years, it increased the ozone problem. Due to a combination of exceptionally cold temperatures in the upper atmosphere and the effect of CFCs the ozone hole over Britain was the deepest ever recorded on the 5 March 1996.

Many amphibian species such as frogs and toads are suffering an unprecedented decline in their numbers over recent years at both tropical and temperate latitudes in places as far apart as Canada and Costa Rica, Australia and Amazonia. This reduction in numbers is a widespread worldwide phenomenon which is causing herpetologists considerable alarm. This is almost certainly a result of some environmental change. Amphibians, with their thin skins, are both extremely susceptible to pollutants and sensitive to ultraviolet radiation which we know is increasing as a result of the ozone hole. Could amphibians be the equivalent of the miners' canaries that are sending us a message about a serious danger?

Fortunately, governments have been fairly responsive to the ozone crisis and the UN Environmental Programme helped to formulate the Montreal Protocol of 1987 which agreed to halve CFC use by 1999. Later amendments in London in 1991 and Copenhagen in 1992 took the matter even more seriously and it was agreed that the industrialized countries should phase out CFCs completely by 1996 and the non-industrialized countries by 2006. However, since CFCs persist for a long time in the atmosphere, it will take almost a hundred years for this ban to take proper effect. CFC substitutes are more expensive than CFCs to produce, but in future we cannot continue to live at the expense of our environment. Countries such as China and India are seeking to install refrigerators in every home and if not assisted by the industrialized world they may continue to use cheap ozone-damaging chemicals. Here is another area in which the industrialized world must provide as much technical assistance as possible and work in partnership with the non-industrialized world if we are to permanently solve the ozone crisis.

Pollution

There are numerous other pollutants besides the greenhouse gases and CFCs, and we have been so careless with our waste products that they are encountered in the air we breathe, in our rivers and oceans, and wherever we go on land. I expected pollution when, out of curiosity, I visited Cubatão, in São Paulo State – at that time (1982) allegedly the most polluted city in the world. I could not believe that humans could live in that atmosphere where it was painful to breathe the cocktail of noxious chemicals mixed with cement dust which floated around the valley. By contrast I remember my surprise, on a visit to Palmer Station in Antarctica, when I encountered scientists working on pollution. They are finding industrial pollutants from São Paulo, Brazil, in the air, and penguins with large quantities of mercury in their bodies in what appeared to be one of the most pristine, beautiful and untouched habitats remaining on earth.

Governments have been comparatively quick to react to some of the most toxic pollutants to health, such as DDT, dioxin and PCBs, the effects of which are quickly seen in industrialized societies, but they have been much slower to control the emissions from cars and factories which belch out huge quantities of carbon dioxide and sulphur and nitrogen oxides.

The US and British governments have both been slow to acknowledge the reality of acid rain, but its existence and its harmful effects can no longer be disputed. 'Forest death' or *Waldsterben* is a familiar term in Germany, where thousands of conifers are dying as a result of acid rain caused by sulphur dioxide and nitrogen oxides, pollutants which form acids and can turn the rain as acid as vinegar. The sterility of the Swedish lakes has been known for several decades and Scandinavian people are constantly pointing to the acidifying effect in their lakes due to coal burnt in Britain and Germany. The effects of acid rain on the forests of the eastern United States and Canada are well documented, and clouds in the eastern United States were more acid than lemon juice when sampled by the New York Botanical Garden Institute of Ecosystems Studies.

A recent report funded by the Forestry Commission and the European Commission and produced by Richard Mather of the Oxford Forestry Institute shows that Britain's trees have been widely damaged by ozone and acid rain over the four years 1989-1992. The worst damage is to beech trees where the upper leaves and branches are dying because of the high ozone concentration. Norwegian spruce is suffering a sharp thinning of the crowns due to both acid rain and sulphur dioxide gas. Although Mather's method heavily favoured explanations other than pollution, he reported that 'there is considerable evidence that trees which tolerate the independent stresses of pollution and drought will exhibit signs of decline under circumstances of combined water and pollution stress'. This combined stress led to the density of foliage in the crowns of beech trees in parts of the West Country and Wales falling from around eighty per cent to sixty per cent between 1989 and 1991. More than eighty per cent of British oak trees had lost a quarter or more of their crown density.[14]

Air pollution is already a costly matter because of its damage to forests, crops, water quality and human health. The expense needed to control emissions and sulphurous gases could be compensated for by the reduction in medical bills and increased agricultural productivity. It has been estimated that acid rain will cost Europe 118 million cubic metres of wood, worth £16 billion annually.[15] It would be better to invest £16 billion in pollution control than to reduce tree growth so drastically.

During the next few decades we need to reduce the emissions of sulphur dioxide and nitrogen oxides considerably through greater use of pollutant-removing scrubbers on factory chimneys, worldwide use of catalytic converters on cars, greater use of cleaner fuels, improved public transport and through acting upon the recommendations of the Royal Commission on Environmental Pollution and Transport to which I have already referred.

Pollution is not confined to acid rain and the atmosphere. Wherever we go, we encounter its effects. Rivers and seas are polluted with sewage and industrial effluents; agricultural land and water tables are polluted by chemical fertilisers, pesticides and radioactive fallout; oceans are polluted by oil slicks and shipwrecked submarines; even outer space is polluted by satellite debris to such an extent that it is a hazard to US Space Shuttle flights.

Recent reminders of the dangers of pollution are the oil spills from the Exxon Valdez in Alaska, the Braer off Shetland and most recently the Sea Empress off the coast of Wales. Two other key events of this decade have drawn attention to pollution: the Gulf War, with the blowing up of hundreds of oil wells; and the opening up of Eastern Europe to reveal the consequences of the complete lack of pollution control by industry. The latter has shown that by comparison the regulatory procedures of the Western industrialized countries have actually had a considerable effect for the good. The next forty years must include a strengthening of the legislative effort to control pollution and considerable aid to Eastern Europe. Lead emissions in the United States have declined impressively as a result of the legislated transfer to unleaded petrol since 1970. At last this is now happening in the United Kingdom. In the United States motorists are paying more to have catalytic converters on their vehicles but still refuse to pay a higher and more reasonable price for their petrol. It is usually cheaper to prevent pollution than to clean it up.

As we in the industrialized world apply this lesson and also clean up our pollution, we must also transfer our technology freely to the non-industrialized world, Eastern Europe and anywhere else that needs assistance. Future costing must include the disposal of waste and the control of pollutants in the price of products. As population inevitably continues to increase over the next forty years, the future of our biosphere will depend upon our ability to control and reduce poisonous emissions.

We have got away with a lot because of nature's extraordinary capacity to act as a sink for unwanted products. Today, we have reached the limits of this natural sink, and in future we must insist that the polluter pays for control of pollutants. Legislation for this is unpopular because of the initial cost and the effect on human lifestyles, but this is a small price to pay for survival.

Biodiversity

Biodiversity has become a buzz word in recent years. It is a term that includes the diversity of species of living organisms on earth, the genes or genetic information which they contain and the complex ecosystems in which they live. Biodiversity is the result of four billion years of evolution. The human race is a recent newcomer that is seriously altering this process. Estimates of the number of species on our planet range from five to fifty million, and many people agree ten million is a conservative estimate.[16] However, only 1.5 million species have so far been named and classified. All species interact with others in intricate ways: the network of predator–prey relationships; the pollination and dispersal of plants by animals; the attack of and the resistance to diseases; the competition for a niche; and in many other ways. Species are not autonomous – they interact with others in an interdependent way and together form the ecosystems of the world, such as tropical rain forests, tundra, heathland, even deserts. Each ecosystem is held together by a delicate web of interrelationships. The whole system can break down when keystone species are removed.

We need to maintain biodiversity into the future for many reasons. The living organisms of the earth maintain our atmosphere and our climate. Without forests and the organisms in the oceans, the life-support system will break down. Yet at present we are still cutting down tropical rainforests at a rate of fifty-four acres per minute. A tropical forest area the size of Kew Gardens disappears every six minutes! In future an *economic* value must be placed on forests, both tropical and temperate, for their role in preserving world climate and acting as a store of carbon. This type of economic thinking is only just beginning.

Humankind depends upon biodiversity for food, medicines, shelter, clothing fibres and industrial products such as rubber, starches and oils, yet at present we are still prepared to allow the extinction of species in such places as Madagascar, Hawaii and Atlantic coastal Brazil. Even at home in Britain extinction is happening to plants and animals. At the Royal Botanic Gardens, Kew we have seeds in our seed bank of at least two species of plants which have become extinct in the wild since we began collecting seeds twenty-three years ago. The interrupted brome grass (*Bromus interruptus*) and the triangular club-rush (*Scirpus triqueter*) would be completely extinct if we had not collected the seeds before their demise in the wild.

Modern agriculture and forestry favours monocultures with little genetic diversity which encourages susceptibility to diseases and pests. The Irish potato famine of the 1840s was caused by the cultivation of a single variety of potato. In contrast we find in the Andes, where the potato is native, any Peruvian market will have twenty to fifty different varieties available. To maintain their crops, plant breeders must return to the wild varieties for disease-resistant species, for new properties such as sweeter tomatoes, or for drought-resistant qualities. As the world climate changes, the need for varieties of crops adapted to different climates, pests and diseases will increase. We now have the technology to transfer genes from one plant to another, but we will have this flexibility only if genetic diversity is preserved.

Humankind seems to have an extraordinary capacity for overusing even the most valuable of natural resources. The population of Easter Island in the Pacific Ocean crashed in the early eighteenth century because they exhausted their forests. There is a tendency to mine rather than sustainably manage our most useful species. In recent years there have been many reports of the collapse of fisheries in different parts of the world especially around North America, Europe and Japan. Despite warnings from scientists for many years, the populations of bottom-dwelling fish such as cod, haddock and flounder are at all-time lows and the fishery industry is collapsing in many places. The US National Marine Fisheries Service estimates that forty-five per cent of fish stocks whose status is known are now overfished and the population of some species is down to ten per cent of the optimum level, the number which would yield the largest sustainable catch. At present it is estimated that New England fishermen catch around sixty per cent of the entire cod population each year which is more than double the sustainable level. The number of spawning cod in the North Sea is down to just five per cent of the figure of twenty years ago. A February 1996 House of Lords report on fisheries condemns the failure of European fisheries ministers, at their meeting in Brussels in December 1995, to make cuts in catches recommended by fisheries scientists. Former chief fisheries advisor of the British government, Michael Holden, said 'the fundamental problem is not the adequacy of the scientific methods and data, but the lack of political will to act upon it'.[17] Self interest and political lobbying are holding back so many necessary controls of our environmental resources. This sort of over exploitation with no thought for the future, exemplified by the over exploitation of fish, can only be the result of human greed.

At Kew we now control pests in our greenhouses by using ten species of insect predators, two fungi and a bacterium. This avoids the use of toxic pesticides. Integrated pest management of this sort is on the increase, but will only be possible if we can preserve the insect species, fungi and bacteria that are needed. In California, the wild brambles harbour a species of wasp which controls a major pest of grapes. It has been estimated that this saves farmers $125 per hectare in pesticide costs.[18] So many wild species, like this wasp and bramble, have an economic or environmental value, yet we continue to destroy them unconcernedly.

Each time a species becomes extinct we narrow our options for the future in some way. Only five per cent of the world's plant species have been properly tested for their medicinal properties. How many potential cures for cancer or AIDS are we losing as extinction progresses? Global extinction rates could already run as high as twenty to fifty species per day, representing an irreplaceable loss of potentially useful natural resources. Human-caused extinction now exceeds the natural rate of extinction by over 10,000 times, and we have entered a phase of mass extinction of species.[19]

We have already lost half of the tropical rain forests of the world which occupy only seven per cent of the land surface yet harbour perhaps seventy per cent of the terrestrial species. Therefore we have little time to save the species that remain. If there is to be a future for the human species, it is essential to preserve all three aspects of biodiversity: large areas of natural habitat which preserve the species and maintain the climatic stability of our planet; as many species as possible because of their roles within ecosystems and because of their potential uses; and as much of their genetic diversity as possible because the future use of many of our crops depends upon the genes of their wild relatives.

The biodiversity issue is absolutely international. The corn crop of the United States depends upon wild species of maize in Mexico for genetic material both to improve crops and to introduce disease resistance. The coffee crop of Brazil depends upon the wild species of the genus *Coffea* in Ethiopia and Madagascar. The world climate depends as much on the forests of the boreal zone in Canada and Siberia as upon the rainforests of Brazil and Zaire. Our future needs more concerted international efforts, without which we are narrowing the options for our descendants and will become known as the generation who diminished biodiversity. It is almost too late, but by turning to sustainable use there is time to reverse the trend and slow down extinction. However, there is no room for complacency. The actions of the next forty years regarding biodiversity will determine whether or not human life will survive on earth. If we continue at the present extinction rate of between 4,000 and 6,000 species a year, by 2030 there will be between 160,000 and 250,000 less species to hold our biosphere together or for us to use or even just enjoy.

Soil

Soil is composed of a mixture of mineral particles, organic material, living microorganisms, water and air and is a resource upon which all humans depend for their food. In a natural ecosystem the soil not only supports the plant life but is replaced through the process of regeneration of the available nutrients and the maintenance of the multitude of organisms that live in it. The natural cover of vegetation prevents erosion of the soil. However, current agricultural practices show little concern for this most precious of resources. It is estimated that the world is losing twenty-five billion tonnes of topsoil a year and India alone loses six billion tonnes a year or almost twenty-five per cent of the world total. However, the erosion rate in Indiana in the corn belt of the USA is the same as that of India, and the United States loses 3 billion tonnes per year.

Soil erosion is not a new problem. Plato in the fourth century BC said that there had been a constant movement of soil away from the high ground and what remained was like a skeleton of a body wasted by disease.

It is taking us a long time to realize that what we are doing to the soil and the extent of the damage today is much more serious than in 400BC. We are not heeding warning signs such as the American dust bowl which occurred in several western states in the 1930s. A recent study by David Pimentel of Cornell University showed that soil erosion is more intense than ever rather than declining.[20] It takes about 500 years to form one inch of topsoil, yet intensive farming methods are leading to topsoil being lost at between twenty and forty times faster than it is being replaced.

Soil erosion is not confined to the major agricultural regions of the world, it is also a serious problem in areas where tropical rainforest has been removed. Most rainforests are over particularly poor soil and when this is exposed by the removal of trees the erosion can be so intense that reforestation is difficult. In some areas where tropical forest has been removed, the process of laterization takes place whereby a rock-hard pebbly soil is produced.

Farming, deforestation and overgrazing have so damaged the soils that eleven per cent of the planet's vegetated surface (an area the size of China and India) has been so badly degraded that it will be either impossible or too costly to restore it. It is absolutely essential to reverse this trend of soil destruction. In the US economic incentives are now being offered to farmers who protect topsoil. As a result soil erosion has dropped by one third between 1985 and 1990 and should do the same again by 1995. This has been achieved by techniques such as conservation tillage where crop residues are left in the soil partially covering it and helping to hold it in place, and by non-tillage agriculture where only small holes for seeds are drilled without disturbing the subsurface soil. Crop rotation, the use of fallow, contour ploughing, strip cropping and terracing are all methods of soil conservation. As individuals we can set an example through composting all our organic waste. I personally use a wormery which is simply a dustbin into which worms are placed to break down organic material. It is a most efficient way of regenerating our organic kitchen waste back into the soil.

Plunder of the Highlands and Islands

Even my beloved Highlands and Islands of Scotland are being affected by human greed. A more remote and beautiful place than the Isle of Harris could not be imagined. Yet it contains resources that are in great demand – sand, gravel and rock aggregates. These are mainly used for roadbuilding. Instead of investing more in public transport, nature reserves and sites of special scientific interest are being destroyed for new roads to be opened up. Suitable materials for road building occur near to most regions where the roads are being built, but quarries and gravel pits are unpopular neighbours and stringent planning controls prevent their operation. A superquarry at Glensanda on the west coast of Scotland sends rock to southeast England, and also exports sixty per cent of its production to Europe and the USA. The Glensanda quarry exports three million tonnes of granite annually. An application has been made to develop another superquarry at Lingerbay, near Rodel on the Isle of Harris. There has been a lengthy public enquiry. The need for aggregate and the resistance to quarries in the south means that unspoiled parts of creation are now being destroyed. The latest news is that another large firm is expected to apply soon for planning permission to open a coastal quarry beside Loch Seaforth, also on Harris. I cannot imagine what my grandmother, who was born in Harris, would have thought of these developments. The age-old granite rocks of that beautiful island are being carted away in bulk containers to destroy other parts of creation around the world and encourage the increase in road traffic. Human greed has no bounds.

Conclusion

In marked contrast to the first chapter where we looked at the goodness, the integrity and the wonder of creation, we have seen here that there is now something seriously wrong with the system and that this is due to the activities of a single species – *Homo sapiens*. Our attitude to the environment is both selfish and has little concern for the future. How unlike the North American woodland Indians we are. They often think in terms of the consequence of their actions upon the seventh unborn generation. Contemporary life is geared to the length of political office.

Speaking about the UK Biodiversity Action Plan, Lord Selborne said, 'Are we all prepared to pay more for domestic heating, motor fuel, road use and Third World products? All political parties know that the promise of economic growth is an essential component of an election manifesto. The challenge is to break the link between economic growth and pollution and to set out a framework for the preservation of habitats and wildlife in a context which is compatible with the legitimate aspirations of the electorate.'[21]

This brief summary of some of the most important aspects of the environment crisis is sufficient to illustrate that there is something seriously wrong with the world today. We live in a world where the rich nations are getting richer and the poor ones poorer; where the number of starving is increasing each year even though we are not yet quite beyond the ability to feed the whole population; where violence abounds and where social injustice continues. Is this a political, sociological and economic crisis or is it a moral and spiritual one? I would maintain that the crisis is spiritual and that it should not be too much of a surprise to the Christian who has been brought up with a belief in the sinful nature of humankind.

The magnitude of the problem demands ethical, moral and religious solutions as well as scientific and technological ones and the challenge to today's church is to respond to this before it is too late. Can we not see the arrogance of what our species is doing to God's creation. Even the United Kingdom's Environmental Strategy (1990) recognizes that it is an ethical crisis: 'The starting point for this Government is the ethical imperative of stewardship which must underline all environmental policies. Mankind has always been capable of great good and evil .... We have a moral duty to look after our planet and to hand it on in good order to future generations.'[22]

Let me conclude with some words of the Prophet Isaiah:

The earth dries up and withers,
The world languishes and withers;
The heavens languish together with the earth.
The earth lies polluted under its inhabitants;
for they have transgressed the laws, violated the statutes,

broken the everlasting covenant ....
The wine dries up, the vine languishes,
all the merry-hearted sigh.
The mirth of the timbrels is stilled.

(Isaiah 24:4-7)

Notes

1 US National Academy, *Population Summit of the World's Scientific Academies, New Delhi, India* (Washington: US National Academy Press, 1993) p.13.

2 IUCN, UNEP, WWF, *Caring for the Earth: A Strategy for Sustainable Living* (Switzerland: Gland, 1991) p.228. (The World Conservation Union, United Nations Environment Programme and Worldwide Fund for Nature)

3 Lawrence Hamilton (ed.), *Ethics, Religion and Biodiversity: Relations between conservation and cultural values* (Cambridge: The White Horse Press, 1993) p.218.

4 This has been shown for deer by D.I. Rasmussen, 'Biotic communities of Kaibab Plateau, Arizona', *Ecological Monographs*, no. 11 (1941) pp.230-75;

for thrips by J. Davidson and H.G. Andrewartha, 'Annual trends in a natural population of *Thrips imaginis* (Thysanoptera)', *Journal of Animal Ecology*, no. 17 (1948) pp.193-99, 200-22;

and for the oscillations between the snowshoe hare and their predator, the lynx, in North America, D.A. MacLulich, 'Fluctuation in numbers of the varying hare (*Lepus americanus*)', *University of Toronto Studies, Biology Series*, no. 43 (1937).

When a predator over exploits its principal source of food its own population also crashes in numbers.

5 Napoleon Chagnon, *Yanomamö: The fierce people* (New York: Holt, Rinehart and Winston,1968).

6 US National Academy, *Population Summit of the World's Scientific Academies, New Delhi, India, op. cit.*

7 American Association of Science report, 1994.

8 Lester Thurow, *Technology Review, Aug./Sept. (1986).*

9 Fred Pearce, 'A breath of fresh air for planet Earth', *New Scientist*, 23 April 1994, pp.16-17.

10 Mick Kelly and Susan Subak, 'How deep must the cuts be?', *Tiempo*, no. 11 (1994) pp. 4-5.

11 *Ibid.*, p.5.

12 Tim Appenzeller, 'Ozone loss hits us where we live', *Science*, vol. 254 (1991) p.645.

13 John Gribbin, 'Ozone low north and south', *New Scientist*, 14 May 1994, p.17.

14 Fred Pearce, 'A breath of fresh air for planet Earth', *op. cit.*

15 William Bown, 'Europe's forests fall to acid rain', *New Scientist*, 11 August 1990, p.17.

16 Paul and Ann Ehrlich, *Extinction: The causes and consequences of the disappearance of species* (New York: Random House, 1981).

    Robert May, 'How many species?', *Philosophical Transactions of the Royal Society of London*, series B.330 (1990) pp.171-82.

    Edward Wilson, *The Diversity of Life* (Harvard University: Belknap Press, 1992).

17 Fred Pearce, 'Only stern words can save the world's fish', *New Scientist*, 10 February 1996, p.4.

18 Office of Technology Assessment, US Congress, *Technologies to Maintain Biological Diversity* (Washington, DC: OTA-f-330 US Government Printing Office, 1987).

19 See, for example, Paul and Ann Ehrlich, *Extinction: The causes and consequences of the disappearance of species, op. cit.*

    Norman Myers, 'Questions of mass extinction', *Biodiversity and Conservation*, no. 2 (1993) pp.2-17.

    Peter Raven, 'The politics of preserving biodiversity', *Bioscience*, no. 40 (1990) pp.769-74.

    Michael Soulé (ed.), *Viable Populations for Conservation*, (New York: Cambridge University Press, 1987).

    Edward Wilson, *The Diversity of Life, op. cit.*

20 David Pimentel, 'Environmental and economic costs of soil erosion and conservation benefits', *Science*, vol. 267 (1995) pp.1117-23.

21 'Biodiversity in action', *Science and Public Affairs*, Spring 1994, pp.56-58.

22 HMSO, 1990

Chapter three

From sin to stewardship

*I brought you into a plentiful land
to eat its fruits and its good things.
But when you entered you defiled my land,
and made my heritage an abomination. (Jeremiah 2:7)*

*The Lord God took the man and put him in
the garden of Eden to till it and keep it. (Genesis 2:15)*

Sin

We have seen in the last chapter that there is something seriously wrong with the world in which we live. This should come as no surprise to the student of the Christian scriptures which, after an account of creation and its goodness, soon changes course drastically and recounts the story of the fall, or the entry of sin into the world. As a result God's garden has become polluted and threatened. Isaiah's words that 'the earth lies polluted under its inhabitants' ring true today.

The fall featured a plant that was so beautiful that it stimulated desire. The fruit of the tree was enticing – 'good for food, and that it was a delight to the eyes' *(Genesis 3:5)*. The tree awakened selfishness, something which God forbade. The temptation was one to disrupt the moral ecology which God had established through the claim of a serpent (be it symbolic or real) that nature could reveal its secrets to humanity in spite of God's wish to withhold them. The serpent suggested that if humans mastered nature, they would be equal to God and be able to shape morality for their own convenience. In fact humanity was called not to be like God where we could manipulate nature to our own advantage, but to function in God's image and as God's stewards. But the fall symbolizes spiritual death or separation from God. From then on people's relationship to both the creator and creation altered radically, and the stewards became the plunderers of nature.

The verses from Isaiah chapter 24 which I have quoted above obviously speak of a moral pollution. This continues today and has followed the history of humankind since the fall. It is moral pollution that causes us to be physical polluters and to abuse the environment. It causes us to forget the goodness and the purpose of creation. Pollution in the biblical sense expresses God's displeasure with the nature of fallen humans and with their adverse impact upon God's creation.

Concurrently with Isaiah preaching in Judah, the prophet Hosea in the Northern Kingdom preached about the way in which human violence can damage nature. Both the land and the sea are mentioned.

Hear the word of the Lord, O people of Israel;
for the Lord has an indictment
against the inhabitants of the land.
There is no faithfulness or loyalty,
and no knowledge of God in the land.
Swearing, lying, and murder,
and stealing and adultery break out;
bloodshed follows bloodshed.
Therefore the land mourns,
and all who live in it languish;
together with the wild animals, and the birds of the air,
even the fish of the sea are perishing.

(Hosea 4:1–3)

Later Hosea prophesied:

For they sow the wind, and they shall reap the whirlwind.
The standing grain has no heads, it shall yield no meal.

(Hosea 8:7)

The worn out soils blowing away today are still fulfilling Hosea's prophecy. Our sinful greed over the soil is now reaping its due reward.

When Jeremiah was instructed to prophesy drought as punishment for the people's sins, in typical fashion he protested to the Lord about the injustice that the land suffered while humans seemed to get away without punishment.

How long will the land mourn,
and the grass of every field wither?
For the wickedness of those who live in it
the animals and the birds are swept away,
and because the people said, 'he is blind to our ways'.

(Jeremiah 12:4)

But it *is* our sinful acts and greed that cause the land to suffer even today. The Lord replied to Jeremiah:

The whole land is made desolate, but no one lays it to heart.

(Jeremiah 12.11)

It is clear that God was pained by the parched earth. It is the sinful nature of humankind that has led us into the environmental crisis which confronts us today. The environmental problems outlined in the last chapter are not just physical ones caused by industrial development. We are facing a moral and spiritual crisis which is at the root of the matter.

Dominion

Then God said,
'Let us make humankind in our image,
according to our likeness;
and let them have dominion over the fish of the sea,
and over the birds of the air,
and over the cattle,
and over all the wild animals of the earth'.

(Genesis 1:26–28)

Dominion, a word which has often been misunderstood, implies caretaking to act as stewards of God's own purposes. It does not, in its biblical sense, imply the establishment of a competing reign, which is what the fall has led to. Dominion is not domination without justice, but rather responsible rule that does not exploit its charges. God gave instructions to share the earth's vegetation with other creatures.

God said, 'See, I have given you every plant yielding seed
that is upon the face of all the earth, and every tree with seed
in its fruit; you shall have them for food. And to every beast
of the earth, and to every bird of the air and to everything
that creeps on the earth, everything that has the breath of
life, I have given every green plant for food.'

(Genesis 1:29–30)

The dominion was not God's authority to use up all the earth's
resources for human needs alone. A problem in the western world
has been that many Christian people have taken God's command
of dominion as a divine authorization to exploit the earth with no
thought for the welfare of other cultures, other creatures, the land-
scape, the mineral resources, the oceans or the atmosphere.

There is no doubt that persuasive and influential misinterpreta-
tions of Christian doctrine have led to environmental destruction
and lack of respect for nature. There is considerable truth in the
allegations of historian Lynn White who in his paper in 1967 in
Science, entitled 'The Historical Roots of Ecological Crisis', lam-
basted the church for its attitude towards the environment.

White was critical of the church, but he also offered a solution
since he called St Francis the patron saint for ecologists because
he was a Christian saint who 'set up a democracy for all God's
creatures'.[1] White's main theme was to blame the current exploitative attitude towards nature on the influence of the church, especially in the Middle Ages. He claimed that Christianity encouraged people's dominance or mastery over nature and that 'Christianity bears a huge burden of guilt' for the exploitative nature of
the environmental crisis. During the last three decades White's
thesis has been much debated, praised and criticized, and it has
spurred some parts of the Christian church into reconsidering
environmental concerns.[2]

Many of White's allegations are true. The church shares the
blame for the exploitation of nature that is so ingrained into our
western society. However, by no means all parts of the church
have endorsed such exploitation and the biblical texts upon which
the church is based never endorses the exploitation of nature for
the selfish benefit of humankind and the destruction of the rest of
life. Today the church is responding to White's challenge by

re-examining the theology of creation, and then reorienting its attitudes and actions. The current environmental crisis is so great that it challenges all major faiths into environmental action. Since the church's theology of creation should lead to a deep respect for nature as God's creation, the church has a strong basis to be involved in environmental issues. The more closely I examine the foundation of the Christian faith in the biblical texts, the more I am convinced that the problem is not a lack of an environmental ethic, but rather an under-emphasis on certain parts of the Judeo-Christian teaching.

Judeo-Christian teaching begins with the Genesis story of how God created the world in six stages. The reiteration of God as creator stresses that the world around us is not ours, *God* created it, and found that it was good, and to be enjoyed and protected. Humankind was given *dominion* over creation. 'Dominion' means 'lordship', implying caretaking, ownership or trusteeship in the spirit of the Hebrew servant kingship, rather than wanton destruction.

This concept is reinforced by verse 15 of the second chapter of Genesis where we read 'The Lord God took the man and put him in the garden of Eden to till it and keep it'. Two interesting commands were given to the first people: to till and to keep the land. The Hebrew word *abad*, translated 'till', also means 'to serve', and the English 'keep' in Hebrew is *shamar*, to watch or preserve. This is quite different from the frequent misinterpretation of what dominion means. We are actually told to serve and preserve the land! The fact that humankind has been given dominion over the rest of creation should lead us to solving rather than creating and exacerbating the environmental problems of today as we execute our trustee responsibility for it.

Lest we should be too introspective and concerned to place the blame for the environmental crisis solely on Christianity, it is perhaps important to look at the rest of the world. Environmental damage occurs under other religions and philosophies. The great Mayan civilization is believed to have collapsed due to overuse of the land and excessive deforestation. We often hear environmentalists paying tribute to oriental religions, but the environmental destruction in eastern countries is as bad as anywhere else. Deforestation has left India and Bangladesh with very little natural forest, Japan has had enormous pollution problems that have only recently been brought under control, and we are only just beginning to hear of the enormity of the environmental problems of China – a country under an anti-religious communist regime! Acid rain is not just a problem for Christian Europe and North America, it is equally a problem for China, Japan and Eastern Europe. Misuse of our resources springs from the folly of fallen humankind, and not from religion or philosophy, although abuse of our environment may be reinforced or modified by a religion's ethic.

Care for the Earth

Further biblical teachings, especially in the five books of the law, contain clear basic guidelines for implementing the original command 'to subdue the earth'. Respect for the land and concern for its welfare lie behind the injunctions of Exodus 23 to leave the land fallow and to observe rest periods. These injunctions are repeated in Leviticus 25 with addition of the Jubilee years. Jahweh states: 'For the land is mine; with me you are but aliens and tenants'. *(Leviticus 25:23)*

The Jubilee called for a redistribution of the land every fiftieth year to eliminate the inequities of ownership. What different places some Latin American countries, where a privileged few people control much of the land, could be with such an ethic.

With the passage of time much of the Judeo-Christian heritage has become anthropocentric, presenting people as the most important part of the world. Surely this is such a travesty of the original richly diverse created order.

The principle of fallow was for the good of all creation:

For six years you shall sow your land and gather in its yield, but the seventh year you shall let it rest and lie fallow, so that the poor of your people may eat; and what they leave the wild animals may eat. You shall do the same with your vineyard, and with your olive orchard.

(Exodus 23:10–11)

The poor, the alien, the animals and the land itself are all to profit from this judicious stewardship of the land. A similar injunction is expressed in the book of Leviticus:

When you reap the harvest of your land, you shall not reap to the very edges of your field or gather the gleanings of your harvest. You shall not strip your vineyard bare, or gather the fallen grapes of your vineyard; you shall leave them for the poor and the alien.

(Leviticus 19:9-10)

How reminiscent to me this is of gifts – the cups of coffee, bananas, water melons, tapioca and other produce – that the Amazonian Indians and local peoples have most generously given to me, a stranger in their midst, when they really could not afford to.

As I read these passages about fallow from the books of the law I think of the experience of the Aymara Indians in the highlands of Bolivia. In that tribe, individual families own areas of land, but it is the community that determines what will be grown for the benefit of all. Each land owner is told what crop to grow, and is regularly instructed to leave an area fallow. This communal organization of agriculture ensures that an adequate diversity of crops, such as potatoes, quinoa and oca, are grown to meet community requirements. It also gives the land time to rest and recuperate. If a family's land is to be fallow for a year, the produce of other members of the tribe will take care of all needs. As a result the Aymara agricultural system has functioned for hundreds of years without destroying the soil. This is a perfect example of how to look after the soil.

Tragically the Bolivian government acted against this system in recent years. It objected that land is 'wasted' and not used during fallow periods. A new, government-imposed system is destroying the soil and breaking down one of the most effective community systems in the world. Hope for the land and the Aymara people is springing from the support of various enlightened Catholic and Protestant missions amongst the Aymara people. The missionaries have seen the contrast between the native system with its strong land ethic, and the imposed, exploitive, capitalist system. In this case the church is supporting the old system of land utilization, and so is protecting land and culture in support of local people's action.

The example of the Aymara Indians is in marked contrast to what has happened in the Indonesian Island of Siberut. This island, off the coast of Sumatra, has been called a tropical paradise because of the long isolation of its fauna, flora and people. It is now at the stage of confrontation with the modern world which seeks to exploit its timber. The influence of Christian mission work has been detrimental to the societal structure of these people who, as animists, originally believed that each object had its own spirit. They believed in an internal harmony in creation, with one religious force known as *kina alau* ('the beyond'), which was concentrated as spirits and souls in the various manifestations of the creation. These original beliefs of this people ensured that they lived in harmony with their environment. An article in the journal *New Scientist* describes this:

> *The manner in which Christianity has been brought to Siberut has had a devastating effect on the island. The traditional religion with its complex set of taboos against the exploitation of nature is now replaced by a bold form of Christianity with no feeling of stewardship, and which ignores the reasons for which traditional beliefs evolved. This has led to a basic change in the economy of the island, with considerably stronger emphasis on producing surplus for sale, clearing more land, gathering more rattan, wearing store-bought cloth, growing the 'more civilized' rice rather than sago and settling down close to a church.[3]*

The article goes on to give further details of how this change in the people of Siberut is ecologically unsound. Here is an opportunity for ecologically aware missionaries to demonstrate the Christian teachings of stewardship rather than the 'bold form' of Christianity that has been brought to the island.

In the case of Siberut, it is the government that is striving for an ecologically sound conservation plan for the island, and they are having to contend with the destructive 'mission' of the church. If 'the earth is the Lord's' *(Psalm 24:1)* what is the role of the religions of the earth – to cherish and nurture or to exploit? Could it be that those peoples who have remained close to the earth are the ones who have something valuable to impart to today's mechanized, profit-oriented so-called Christian society?

I remember my first day of botanical field work in the rainforests of Surinam. After a long journey by air, two days by canoe and two on foot, we reached the area where our expedition was already at work. Near to our camp was a huge flowering tree of *Licania* in the plant family *Chrysobalanaceae* on which I had recently completed my doctoral thesis. Since the tree was far too large to climb, the expedition leader immediately asked one of our most willing Djika helpers to cut down this forest giant in order for us to collect a few flowering twigs for specimens. The Djika people are descended from escaped African slaves, who reformed a tribal life in the forest of Surinam. They are a wonderful people who have maintained much of their ancient culture.

We were all surprised when Frederik, who was our expedition cook, refused to cut down the tree. After much coaxing he agreed to cut it down, not immediately, but, in a few minutes' time, and only after he had appeased the Bushy Mama, his god. After a chanting prayer and an offering to Bushy Mama, Frederik began to cut down the tree. As he cut he sang. It was a loud chant that would show clearly that it was not he who was cutting the tree, but that it was the white man who should be blamed for this unnecessary destruction. I have never forgotten this experience with Frederik on my first day in the rainforest. It was my first lesson in respecting the beliefs of indigenous peoples. It also changed my plant collecting habits on subsequent expeditions when I was in charge. Instead of felling trees, I have taken along tree climbers to obtain specimens of the leaves and flowers, or I have used long extendible aluminium poles with tree pruners on the end to reach far around the crown of a tree. Botanists only fell a few trees but they must not become as destructive as the missionaries of Siberut.

When these wise Djika people must fell a tree for housing or for their splendid wood carvings, it is done with respect and appeasement to the Bushy Mama. For them the trees of the forest have a spiritual value. The forest is not for careless use but is the possession of its creator.

Before I visited the Maku Indians on the upper Rio Negro in Brazil, I had read about their fish poisoning festivals where they catch a huge number of fish, using plants which stun them, and have a large feast. After spending a few days with the Maku I finally persuaded the chief to allow us to take part in such a fishing expedition. He said it would not be in the nearby stream but another one a short walk through the forest. After a two-hour walk, which at the Indians' pace is a jog, we came to a stream which appeared ideal for fishing, but the chief said it was not this one. Two hours later the next inviting stream was crossed, but again the fishing was not to be there. In spite of their heavy loads of the particular plant they were going to use for fishing, the Indians trotted on through the forest until eight hours later we reached the designated spot on a stream that looked exactly like the four or five which we had already crossed. The botanists were exhausted, but the Maku immediately set about the business of building a small log bridge over the stream, and placing their leaves of the spurge plant, *Euphorbia cotinifolia,* on it. This plant has chemicals which react with the gill-membranes of fish and causes them to become asphyxiated and float to the surface. While the men beat the leaves with sticks to allow the sap from the plant to fall into the river, some of the women stirred up the mud upstream, whilst others began to gather up the stunned fish downstream. Gathering up one's catch in this way is easy!

They collected a huge amount of fish reminiscent of the bounty when the resurrected Jesus advised his disciples to cast their net on the other side of their boat: 'He [Jesus] said to them, "Cast the net to the right side of the boat, and you will find some [fish]." So they cast it, and now they were not able to haul it in because there were so many fish'. *(John 21:6)* In our case a great feast followed. However, I did not recount this story simply to compare it with a biblical one. When I asked Chief Joaquim why we could not have fished at the first stream we crossed, he replied that they had fished there a few months ago, and that each of the other streams we crossed had been fished within the period of the last twenty moons. He said, 'If we fish like this in a river more than once in twenty moons there will be no fish left. They need time to breed again.' How unlike our current world situation where in one place after another the supplies are running out due to over-fishing and the techniques used. Here in the upper reaches of the Amazon the local people are much better caretakers of creation than most of the rest of us.

The examples I have given of indigenous peoples such as the Aymara of Bolivia, the Maku of Brazil and Frederik the cook in Surinam contrast markedly with the materialistic culture which has been thrust upon the people of Siberut to change them from stewards to destroyers of their island paradise. The environmental degradation is merely a symptom of the greed and sinful nature of humankind, especially our western industrialized and materialistic society. This has led to much social injustice and inequity of resources and the oppression of the very people from whom we could learn so much, these indigenous, tribal people who are still close to the land. There cannot be any hope for wise stewardship of creation if we do not address the issues of social injustice. As a conservationist I know that to conserve the biodiversity of the world will require a lot more than addressing the major environmental problems to which I referred in the last chapter.

In the industrialized world we have an affluent lifestyle, which is having a deleterious effect on the environment *and* on the peoples of the non-industrialized world. The distribution of resources is most inequitable. The industrialized countries contain about twenty-two per cent of the world's population but control about eighty-five per cent of its finances. The seventy-eight per cent of the world's people who live in the non-industrialized countries use only twenty per cent of the world's industrial energy. The United States, with only 4.5 per cent of the world's population, produces one quarter of the world's carbon dioxide emissions. Our level of consumption of energy in the industrialized world means that each individual is much more costly to the environment than an individual from a poor country. One average British person consumes over one hundred times more energy per year than one Bangladeshi. But the inequity does not end there. We in the industrialized world consume seventy per cent of the world's grain to sustain over twenty-three per cent of the population. Much of this grain is grown to feed beef cattle, while 1.2 billion people in the world live in absolute poverty and face starvation. In contrast to this, one third of the food purchased in the United Kingdom ends up in the dustbin. It is hardly surprising that whilst the industrialized nations criticized the non-industrialized world about population increase at the Cairo population summit, the latter nations retorted by berating us for our excessive consumption of resources!

One of the worst inequities of all is the way in which the industrialized world has put so many of the non-industrialized countries into debt. Many countries spend a large proportion of their resources merely servicing the interest on their loans and not even repaying the debt. Over half of Ghana's foreign exchange is spent on repaying loans. As a result debtor countries have no way of combating starvation and poverty let alone addressing issues of environmental conservation.

The Bible leaves us in no doubt that we should be addressing these issues. Jesus said:

> *Come you that are blessed by my Father, inherit the king-dom prepared for you from the foundation of the world; for I was hungry and you gave me food, I was thirsty and you gave me something to drink, I was a stranger and you welcomed me, I was naked and you gave me clothing, I was sick and you took care of me, I was in prison and you visited me. Then the righteous will answer him, 'Lord when was it that we saw you hungry and gave you food, or thirsty and gave you something to drink? And when was it that we saw you a stranger and welcomed you, or naked and gave you clothing? And when was it that we saw you sick or in prison and visited you?' And the King will answer them, 'Truly I tell you, just as you did it to one of the least of those who are members of my family, you did it to me'. Then he will say to those at his left hand, 'You that are accursed, depart from me, into eternal fire prepared for the devil and his angels; for I was hungry and you gave me no food, I was thirsty and you gave me nothing to drink, I was a stranger and you did not welcome me, naked and you did not give me clothing, sick and in prison and you did not visit me'.*
>
> *(Matthew 25:34-43)*

Those are well-known words which need no comment, but the Bible is full of admonitions to take care of the poor and needy and to act for social justice. Christ was probably echoing the words of Isaiah in answering the complaint of his people as to why God was rejecting their fasting and worship since the prophet said:

> *Look, you serve your own interest on your fast day,*
> *and oppress all your workers ....*
> *Is not this the fast that I choose,*
> *to loose the bonds of injustice,*
> *to undo the thongs of the yoke,*
> *to let the oppressed go free,*

and to break every yoke?
Is it not to share your bread with the hungry,
and bring the homeless poor into your house;
when you see the naked, to cover them,
and not to hide yourself from your own kin?

(Isaiah 58:3, 6-7)

I have covered a lot of ground in this chapter, from the garden of Eden to the Amazon rainforest, from the wise management of soil by the Amazon Indians to the destruction of the paradise island of Siberut by Christian missionaries and commercial developers, and from environmental matters to those of social justice. But these different topics are linked together by the desperate need for humankind to turn from destroying creation to using it sustainably, and to become part of it again rather than live as a disconnected predator. We have seen that people who live nearer to the natural environment often take better care of it.

My reaction as a young Christian to the experience with Frederick the cook and many other incidences which show care for creation by indigenous peoples could have been different. I could have embraced their animist religion and rejected my own, or alternatively, I could see what Christianity had to offer about caring for creation. I do not place indigenous peoples on a pedestal. Not all tribal people are good conservers of their environment. However, when native peoples do protect the environment or use it in sustainable ways it is usually connected with their taboos and religious beliefs. I believe that advocates of the New Age movement take up these animist beliefs in a pantheistic religion which accumulates the good points of many different traditions, yet offers little hope since the New Agers worship creation and not the creator. Several churches at which I have spoken in recent years have expressed concern about the loss of their young people to the New Age movement. This is perhaps because the church has neglected the rich teachings on care for creation and social justice that are to be found throughout the Bible. It is vital, not only for nature, but for the future health of the church itself, that we have a well developed theology of creation and of Christian stewardship. If there is to be any hope (the theme of the next chapter) then the church must give a lead because it has so much to offer in its rich theology of creation and stewardship.

True Christianity is not only about personal salvation and a place in heaven, but it is also a faith that renews sinful people and sets us on the pathway from sin to stewardship of creation and social justice. Our faith can lead us on, from the sort of greedy exploitation outlined in the last chapter, to the nurture of the part of creation that is around us.

Notes

1   Lynn White, 'The historical roots of our ecological crisis', *Science*, March 1967, pp.1203-07.

2   A good rebuttal to White's paper is that of R. Hiers, 'Ecology, biblical theology, and methodology: biblical perspectives on the environment', *Zygon*, no. 19 (1984) pp.43–60.

3   A. Whitten and Z. Sardar, 'Master plan for a tropical paradise', *New Scientist*, no. 93, 23 July 1981, pp.230–35.

Chapter four

Is there any hope?

The creation itself will be set free
from its bondage to decay and will obtain
the freedom of the glory of the children of God.

(Romans 8:21)

Christianity is a faith of hope and one of the hopes for the future of creation is that Christians might reflect better the image of God in which we were created. Christ died not just for humankind but for the whole world and so making it a better place and being caretakers of it, until he returns, is certainly part of our Christian responsibility. As St Paul said, 'A little yeast leavens the whole batch of dough' *(Galatians 5:9)*. We were called by Christ to be the salt of the earth and so, if there is to be any hope for creation, it will depend upon God's servants getting out there and doing something about it. I hope that our new life through our relationship to Jesus empowers us with a vision that can bring harmony between humankind and nature.

One thousand five hundred miles up the Amazon river a seed of hope is germinating. The rubber tappers of the Brazilian states of Acre and Rondônia are mobilizing. Acre and Rondônia are the two states in which deforestation has accelerated recently, and forests are being replaced by cattle pasture and cacao plantations. Traditionally, these forests have been some of the most productive in terms of extraction of rubber and Brazil nuts. The gathering of both these products does no harm to the forest. Brazil nuts are produced in the fruit of a tree and when they fall to the ground they are simply gathered from the forest floor. The rubber tappers skilfully slit the bark of the rubber trees deep enough for the milky sap to run out into their small cups, but not deep enough to kill the trees. This sort of crop from the forest is ideal because its extraction does not harm the forest and is, therefore, sustainable. Faced with the loss of their land, their forests and their livelihood, the rubber gatherers have formed a union to defend their rights.

The Brazilian priest and liberation theologist, Clodovis Boff, in his reflective journal *Feet-on-the-ground Theology* summarizes the basis of their mobilization when he writes:

'Feet-on-the-ground' in the first place, because this is a theology moving along over the fertile earth. An earthy theology, like a stretch of land pregnant within the seeds of future life.

'Feet-on-the-ground' secondly because this theology is worked out first with the feet. This sort of theological thinking starts with the feet, moves through the whole body, and rises to the head. There are some things you can grasp only by going there and seeing for yourself. This theology says what it has seen and heard as it moved about in the midst of people.

Finally, 'feet-on-the-ground' means that it takes into account the life of those who go around with their feet on the ground. Of those who live on the rock bottom of history, the poor and the oppressed. Those who have been knocked down on the ground, but who keep getting up. A theology of the poor, worked out with them, one that is theirs.[1]

In practical terms this translates into a coalition of rubber tappers and theologians which succeeded in pressuring the State of Acre to set up a series of extraction reserves. These are areas of forest where deforestation is prohibited by state law, but the local residents are allowed to use them for the extraction of forest products such as rubber and Brazil nuts. The first extraction reserve, supporting seven hundred families, was set up in February 1988. Two more followed later that year, and now a whole series have been set up in several different Brazilian states. They have been one of the ways in which the rate of deforestation in the Amazon region has been slowed down.

The Amazon rainforest can be used to grow a variety of products, and these can be extracted year after year instead of cutting down the forest. This is the way in which indigenous peoples have used the forest for many generations. The gathering of nuts or the tapping of trees for latex does not kill the trees, and there are other products such as fruit, basket-making fibre and medicines that can also be harvested from an extraction forest.

Woman working in a nut factory in Brazil where the working conditions are harsh and pay is minimal.
Photo: Ghillean Prance

Woman working in the rainforest, breaking open babussu palm fruit to extract oil, harming neither the forest nor herself.
Photo: Ghillean Prance

Recent studies have even shown that the income yielded from extraction products from any unit area of rainforest is higher than income from a similar area converted to cattle pasture or timber plantations.[2] The concept of the extraction forest is both economically and ecologically more viable than most alternative plans for tropical rainforest.

Why has this radical coalition of rubber tappers and priests been necessary? The conflict for the land in most of Latin America is appalling and is the cause of much unrest. In Brazil where 1.2 per cent of the nation's farmers control sixty per cent of the available land it is not surprising that there is conflict. In 1985 two rural workers were killed every three days by the nation's larger landowners and this rose to one per day in 1986. In 1988 the leader of the rubber tappers union of Acre, Chico Mendes, was gunned down, and the conflict continues today as the landless seek some land.

Why is this rubber cutters' union unusual? Because here a major religious faith has become a catalyst for a stance linking a local environmental crisis with a practical and viable solution.

It is tragically strange that faith and environmental ethics have been sacrificed in the name of progress, greed, development or ridding a culture of traces of so-called paganism. But are the rubber tappers of Acre or the Aymara of Bolivia a ray of hope? In both cases it is the church that has helped to improve their lot: in the case of the Aymara to restore a traditional way of soil management and, therefore, their agricultural system; in the case of the rubber tappers, their right to get a living from the forest and to challenge the developers who would cut down the trees and destroy the rainforest forever.

It is a wonderful experience to sit in at a meeting of a base community such as that of the rubber tappers. Here is a group of uneducated people earnestly studying the Bible and praying to work out what their plan of action should be. Some of them have learned to read as adults so that they can understand the Bible for themselves and use it to influence their decisions.

This is true New Testament Christianity reminiscent of what it must have been like at the time of the early church in Jerusalem as early Christians sought the way forward. It is interesting and important to note how social justice was so much a part of that community.

There was not a needy person among them, for as many as owned lands or houses sold them, and brought the proceeds of what was sold. They laid it at the apostle's feet, and it was distributed to each as any had need.

(Acts 4:34-35)

The pathway of those whose contemporary faith has led them this way is by no means easy. There have been many martyrs amongst the rubber tappers and the attitude of many Latin American governments has been to label these Christians communists. One of their leaders, the saintly former Archbishop of Recife and Olinda in the poverty-stricken northeast Brazil, has done more for the poor than anyone else in his country. He sold the churches' jewels to feed the poor and lives in a small slum dwelling rather than his bishop's palace. One of his most telling statements is:

When I feed the poor you call me a saint;
when I ask why the poor have no food
you call me a communist.

Yet no significant action of these people is taken without the consultation of the scriptures. What then would they find about creation in the New Testament, in addition to what they find about social justice? We have already seen that the Old Testament has plenty to say about both the stewardship of the earth and about creation. The New Testament does not have any instructions about care for the earth, but it does have plenty to say about our relationship to creation.

In Paul's letter to the Roman church we have some important verses which tell us how the harmony of the universe was broken, yet at the same time bringing hope.

For the creation waits with eager longing for the revealing of the children of God; for the creation was subjected to futility, not of its own will, but by the will of the one who subjected it, in hope that the creation itself will be set free from its bondage to decay and will obtain the freedom of the glory of the children of God. We know that the whole creation has been groaning in labor pains until now; and now not only the creation, but we ourselves, who have the first fruits of the Spirit, groan inwardly while we wait for adoption, the redemption of our bodies. For in hope we were saved.

(Romans 8:19-24)

These verses took on a new meaning for me when I read the paraphrase of them by Paulos Mar Gregorios, the Metropolitan of Delhi in the Indian Orthodox Church, a church that has always kept close to creation.

For I regard the troubles that befall us in this present time as trivial when compared with the magnificent goodness of God that is to be manifested in us. For the created order awaits, with eager longing, with neck outstretched, the full manifestation of the children of God. The futility or emptiness to which the created order is now subject is not something intrinsic to it. The Creator made the creation contingent, in his ordering, upon hope; for the creation itself has something to look forward to – namely, to be freed from its present enslavement to disintegration. The creation itself is to share in the freedom, in the glorious and undying goodness, of the children of God. For we know how the whole creation up till now has been groaning together in agony, in a common pain. And not just the non-human created order – even we ourselves, as Christians, who have received the advanced gift of the Holy Spirit, are now groaning within ourselves, for we are also waiting – waiting for the transformation of our bodies and for full experiencing of our adoption as God's children. For it is by that waiting with hope that we are being saved today. We do not hope for something which we already see. Once one sees something, there is no point in continuing to hope to see it. What we hope for is what we have not yet seen; we await its manifestation with patient endurance.[3]

The creation was subjected to 'futility' and 'emptiness' because the perfect harmony was broken by the fall. The doctrine of the fall instructs us that the universe is no longer what God created it to be. Since the fall the command to have dominion has been abused and humanity has been heading towards ecological disaster. What we see around us today can all be traced back to the origin of sin and to the sin that is inborn in humankind. But the good news is that redemption is for all of creation. Humans are redeemed but non-human creation is set free from the bondage it has suffered under the weight of evil.

Christ did not come only to rescue a few believers in Him from this world. He came to renew creation, to restore humanity and nature to full communion with God. The following two quotations from St Paul echo the beginning of John's Gospel. 'In the beginning was the word' *(John 1:1)*. Here Jesus is called the word *(logos)* who was with God at the beginning and through Him all things were made. It follows that Reconciliation to God in Christ reconciles us to creation and also to responsibility for it. One of the most quoted Christian Scriptures from the Gospel of John – 'For God so loved the world that he gave His only son' *(3.16)* – does not say that God loved the people, but rather the *kosmos*, or the whole world. The world to which Christ came was not just that of humans but to the whole created universe. The Christian is someone who is reconciled to the whole of creation through the work of Christ.

For the Christian the 'subduing and utilization' of creation is to be accomplished in obedience to the biblical declaration that God is the ultimate owner and trustee. The church as a whole has been slow to accept this responsibility of stewardship of creation.

Important to the Christian is Christ's place in creation and his redemption of it. For example, in Colossians St Paul writes:

He is the image of the invisible God, the first-born of all creation; for in Him all things were created, things visible and invisible, whether thrones or dominions or rulers or powers – all things have been created through him and for him. He himself is before all things, and in him all things hold together.

(Colossians 1:15-17)

Listen again to Paulos Mar Gregorios:

He, Christ, the Beloved Son, is the manifest presence of the unmanifest God. He is the Elder Brother of all things created, for it was by him and in him that all things were created, whether here on earth in the sensible world or in the world beyond the horizon of your senses which we call heaven. Even institutions like royal thrones, seats of lords and rulers – all forms of authority. All things were created through him, by him, in him. But he himself is before all things; in him they consist and subsist. He is the head of the body, the Church. He is the New Beginning, the Firstborn from the dead; thus he becomes in all respects pre-eminent. For it was [God's] good pleasure that in Christ all fullness should dwell; it is through him and in him that all things are to be reconciled and reharmonized. For he has removed the contradiction and made peace by his own blood.[4]

Here Christ's work is clearly related both to humans and to the non-human elements of Creation, to *all* things. The wonder of redemption through Christ is that it extends far beyond the individual person to all of creation. St Paul also wrote on a similar theme in his letter to the Ephesians:

With all wisdom and insight he has made known to us the mystery of his will, according to his good pleasure that he set forth in Christ, as a plan for the fullness of time, to gather up all things in him, things in heaven and things on earth.

(Ephesians 1:8-10)

To end this list of New Testament quotes I conclude with a few words that start the letter to the Hebrew Christians:

Long ago God spoke to our ancestors in many and various ways by the prophets, but in these last days he has spoken to us by a Son, whom he appointed heir of all things, through whom he also created the worlds. He is the reflection of God's glory and the exact imprint of God's very being, and he sustains all things by his powerful word.

(Hebrews 1:1-3)

If there is hope in Jesus then there is hope for creation, for nature because it has been redeemed through Christ. But, for now, this depends upon those who know the creator being responsible caretakers of creation.

In the face of exploitation, pride, greed and misinterpretation, can it be that there are any signs of hope? Can the religious forces of the world be harvested to restore the earth?

There are the rubber tappers of Acre, the Aymaras of Bolivia, and many other small sparks of hope. Was it in part the persistence and faith of such people that spurred new and exciting links between religion and the environmental crisis?

Mobilization of religious action for the environment was stimulated by the Worldwide Fund for Nature International as part of its celebrations of its twenty-fifth anniversary in 1986. The Fund challenged the great religions of the world to 'create a credibly philosophy for the next quarter of a century of its existence'. This was the theme of the opening address by HRH Prince Philip at an historic meeting which took place, appropriately, in Assisi, the home of St Francis, one of the first Christian conservationists. The meeting in Assisi produced a series of declarations on people and nature, from Buddhism, Christianity, Hinduism, Islam and Judaism.[5] Representatives of each faith outlined how their faith led them to a conservation ethic. The declarations challenge many aspects of secular conservation, and stress that it is not necessarily motives which prioritize the survival of people that should guide our conservation ethic. The whole earth and its creatures should be taken into account. This initiative led to the formation of an international network of conservation organizations and religions which has produced some useful publications. It has also lead to some New Age type of festivals in various churches and cathedrals. I find these problematic and question whether Christians can worship together with those who believe in more than one god.

Until recently, Christian theologians have rarely been greatly concerned about environmental issues. This is in marked contrast to those of some of the other major religions of the world, especially the Buddhist faith. However, today, when faced by a severe environmental crisis and such critics as Lynn White, more and more Christians of all denominations are realizing that a conservation ethic must be an integral part of their faith.

Within the Christian church today there are many signs of hope. Institutes such as the Au Sable Institute of Environmental Studies in Mancelona, Michigan is devoting its entire effort to teaching environmental stewardship and to promoting dialogue among concerned Christians and environmentalists. The North American Conference on Christianity and Ecology (NACCE), whose first meeting attracted over 500 people in 1988, is a good example of an interdenominational approach to the issues of environment.[6] Since 1988 NACCE has issued the useful publication *Firmament*, a journal of Christian ecology. NACCE's mission is to encourage churches to become centres of creation awareness and to teach reverence for God's creation.

The first encyclical letter sent by the pope to all Roman Catholic bishops on ecology was entitled '*Sollicitudo Rei Sociales*' (Anxiety About Social Issues). It instructs that 'one cannot use with impunity the different categories of being, whether living or inanimate, animals, plants, the natural elements, simply as one wishes, according to one's economic needs'.

In 1987 the United Nations Environment Programme initiated an Environmental Sabbath in the United States to encourage the different faiths towards greater stewardship of the earth. This has been repeated and expanded since then.

In the UK we have Christian Ecology Link (CEL)[7] which brings together environmentally concerned Christians and publishes the journal *Green Christians* which is intended as a forum to reflect and contribute to current thinking on faith and ecology issues. CEL also produces many other useful resources such as an education pack for churches entitled *Steps Towards Sustainability*.

1994 saw the first issue of a new quarterly magazine *Green Cross*, an interdenominational Christian publication to promote the care of creation in a way that is faithful to Jesus Christ. There is also the Religious Education and Environment Programme which exists to facilitate teacher training on environmental and spiritual values through participating workshops. One of the key ways of arousing concern must be through education of children.

Clearly the action has begun. If the Christian church and the other major faiths of the world were to mobilize along the lines of their statements in the Assisi declaration we could change the world's environmental situation in a short time. There is a challenge to all people of faith to become grounded in our environmental ethics and to act within our faith communities.

As Christians we need to follow the example of the Amazon rubber tappers and go back to the basic foundation for our faith, the Bible. May we each work out a sound creation theology that acknowledges both the cause of the environmental and social justice problems – the fall – and is based firmly on our redemption and the redemption of creation through the work of Jesus Christ. As Christians we have a special responsibility to be stewards of the environment because it is God's good creation.

To achieve this we need the help of the Holy Spirit, that precious wind, breath or air *(ruach)* of God that swept over the face of the waters *(Genesis 1:2)* during creation, the wind that blew over the waters and calmed them *(Genesis 8)* to renew creation after the flood. It was the wind that blew the waters apart and then onto the Egyptians as they tried to stop the Israelites from departing on the journey towards the promised land *(Exodus 15:10)*. Recently I was walking on the moors of the Highlands of Scotland breathing some of the purest air possible, pleasantly scented with the aroma of heather and pines. As we pollute the air and poison ourselves, we are destroying creation, corrupting the very spirit by which we live. Our physical and our spiritual survival are closely connected to the air that we breath and to the work of God's breath of fresh air, the spirit.

The wind blows where it chooses, and you hear the sound of it, but you do not know where it comes from or where it goes. So it is with everyone who is born of the Spirit.

(John 3:8)

Notes

1   Clodovis Boff, *Feet on the Ground Theology: A Brazilian journey* (Maryknoll, New York: Orbis Books, 1987) pp.xi, xii.

2   Michael Balick and R. Mendelsohn, 'Assessing the economic value of traditional medicines from tropical rainforests', *Conservation Biology*, no. 6 (1992) pp.128–30.

Charles Peters, Alwyn Gentry and R. Mendelsohn, 'Valuation of an Amazonian rainforest', *Nature*, no. 339 (1989) pp.655–56.

3   Romans 8:18-25, translated by Paulos Mar Gregorios in Wesley Granberg Michaelson (ed.), *Tending the Garden: Essays on the gospel and the earth* (Grand Rapids, Michigan: William B. Eerdmans Publishing Company, 1987) p.84.

4   *Ibid.*, p.87.

5   For details see Worldwide Fund for Nature, *The Assisi Declarations: Messages on man and nature from Buddhism, Christianity, Hinduism, Islam and Judaism* (Switzerland: Gland, 1986). Later they were added to by the Baha'i.

6   For the proceedings of this conference see Frederick Krueger (ed.), *Christian Ecology: Building an environmental ethic for the twenty-first century* (San Francisco: NACCE, 1988).

7   Christian Ecology Link, *Steps Towards Sustainability* (Harrogate: Christian Ecology Link, 1994).

Chapter five

Actions for an earthkeeping church

*To accept God as the Creator of all things, implies that
man's own creative activity should be in cooperation with
the purposes of the Creator who has made all things good.*[1]

(Church of England Doctrine Commission)

It is no good simply listening to and agreeing with the theology
of creation. What is needed in the church is action by its members
and so I conclude with a few suggestions as to how we can be
better caretakers of creation both as individuals and through our
churches. Let us remember that Christians have a special responsibility to the environment because of our relationship with the
creator. We are called to examine our lifestyles, to be the stewards
of creation and to fight for social justice. To bring *shalom* or
'wholeness and completeness' to the world around us. We *can* be
the yeast that leavens the dough if we follow a few guidelines.

Actions for individuals

1   Reduce your energy consumption to save fossil fuels and re-
    duce the greenhouse effect.

–   Use public transport, walk or cycle whenever possible.

–   Switch off all unneeded lights.

–   Improve your home insulation and seal off draughty places
    around doors and windows. Use double glazing using low-
    emissivity glass if you can afford it.

–   Lag your hot water tank.

–   Heat and cook with gas which is much more efficient than
    electricity.

–   Use low-energy light bulbs whenever possible. They use
    twenty per cent of the energy of incandescent bulbs and last
    eight times longer.

2   Eat environmentally sound food.
–   Buy fresh, locally produced food that has not been transported over long distances or grow your own organically.
–   Avoid highly processed food.
–   Eat less meat.

3   Recycle your waste, especially paper, glass, plastic and clothing, and buy recycled products.

4   Buy products from sustainable and equitable sources. Use a green consumers guide for advice on purchases and avoid products that have caused environmental damage or exploited people. (For reference, Richard Adams, *Shopping For a Better World*.)

5   Pray for justice, peace and wise use of resources and for the church to promote more stewardship of our planet.

6   Write letters on local and international issues to your MP, MEP and council representatives, and vote along environmental lines.

7   Garden for the environment
–   Produce your own vegetables, fruit and herbs using organic means.
–   Compost all your household and organic garden waste. A wormery is an excellent way of using your household waste to produce both liquid fertiliser and compost.
–   Avoid the use of peat products in your garden because they are taken from bogs, destroying this habitat and endangering various species.
–   Collect rainwater for watering your plants.

8   Join a local or national environmental group, learn from it and take an active part in their programme.

9   Join Christian Ecology Link, 20 Carlton Road, Harrogate, HG2 8DD and use their resource materials.

10 Use ethical investment trusts. Do not invest in companies that destroy the environment or in banks that have large third-world debtors. (CEL issues a list of ethical investment trusts).

11 Question your expenditure on luxury items and resist materialism.

12 Avoid excess wrapping and take a shopping bag to the market to avoid using extra disposable ones.

13 Make sure your car runs on unleaded fuel. When it is replaced check that it has a catalytic converter and that it is an economic consumer of fuel.

Actions for churches

1   Share transport to church to save fuel and, if possible, join a church to which you can walk.

2   Carry out an environmental audit of your church. Is it well insulated? Does it use low energy light bulbs? (CEL produces a pamphlet about an ecological audit for a church.)

3   Could your church plan recycling skips on its property, perhaps in its car park?

4   Ensure that you have regular acts of worship with an environmental theme. The harvest festival is an ideal time to remember the environment.

5   Have a church tree-planting day on church property or elsewhere.

6　Collect used tools or other useful objects for refurbishing and distribution to the non-industrialized world.

7　Encourage generous giving to organizations such as CARE, CAFOD and the A Rocha Trust which are addressing many of the issues of poverty, injustice and the environment.

8　Make sure that your church contacts local authorities, MPs and MEPs about environmental issues.

9　If your church has funds to invest, make sure that they are invested ethically.

10　Use your churchyard for conservation purposes and manage it in an ecologically sound way. Churches own a lot of property that could set an example. The Church and Conservation Project of the Arthur Rank Centre, Stoneleigh, provides information on managing church grounds to encourage wildlife.

I will end with some wise and hopeful words of Bishop Hugh Montefiore who chaired one of the lectures upon which this text is based.

The Christian view of history involves the concept of purpose and design. Saint Paul looks forward to the time when God will be 'all in all'. Revelation looks forward to the new creation, a new heaven and a new earth. All will be recapitulated. Nothing will be wasted. Redemption is in Christian theology, to be extended to the whole divine purpose for the world of nature, bringing to birth new possibilities at present unknown and enabling the whole world to reach the fulfilment of its potentialities.[2]

Notes

1　Hugh Montefiore, *Man and Nature* (London: Collins, 1975).

2　Hugh Montefiore, 'Ecology, theology and posterity', *New Scientist and Science Journal*, no. 49 (1971) pp.316–18.

Useful resources

The Biblical Call for Earthkeeping
Video issued by Justice and Care of the Earth Task Force, South-Central Synod of Wisconsin, Evangelical Lutheran Church in America. (Available only in VHS). Available from Au Sable Institute of Environmental Studies, 7526 Sunset Trail NE, Mancelona, MI 49659, USA. $17.95

Steps Towards Sustainability
Resource pack from Christian Ecology Link, 20 Carlton Road, Harrogate HG2 8DD. £4 including postage.

The Living Churchyard
Resource pack from the Church and Conservation Project, The Arthur Rank Centre, Stoneleigh Park, Warwickshire CV8 2LZ. Information on managing churchyards to encourage wildlife. £6.00

How Green is Your Church?
A fifteen-minute video available from CEL Publications, 20 Carton Road, Harrogate HG2 8DD. £15.00

The Healing of Creation
The report of the Maranatha Commission. The Maranatha Community, 102 Irlam Road, Manchester M41 6JT, 1994. £2.00

Evangelical Environmental Newsletter
Editor Dr R. C. Carling, Christian Impact, St. Peter's Church, Vere Street, London W1M 9HP. £10 for a one-year subscription; £18 for a two-year subscription.

The Environmental Prayer Letter
Available from Philip Clarkson Webb, 15 Valley View, Southborough, Tunbridge Wells, Kent TN4 0SY. Annual subscription £6.00.

The Bible and God's Environment
A set of four Bible studies. Published by the Board for Mission and Social Responsibility, 2778 East Park Road, Leicester LE5 5AY.

Floods and Rainbows: A study guide on the environment for those who care about the future
(London: Division of Social Responsibility, Methodist Church, 1991) £4.00

The Society, Religion and Technology Project of the Church of Scoland
John Knox House, 45 High Street, Edinburgh EH1 1SR.

A Rocha Trust
Christians in conservation. Their work includes a bird sanctuary and Christian research station in Portugal and many other activities linking Christians in practical conservation. Contact 3 Hooper Street, Cambridge CB11 2NZ.

International Evangelical Environmental Network
Contact Dr Chris Sugden, Oxford Centre for Mission Studies, St Philip and St James Church, Woodstock Road, Oxford OX2 6HB.

Further reading

Richard Adams et al, *Shopping for a Better World*, (London: Kogan Page, 1991).

Peter Bakken et al, *Ecology, Justice and Christian Faith: A critical guide to the literature* (Westport, USA: Greenwood Publishing, 1995).

Wendell Berry, *The Gift of Good Land: Further essays, cultural and agricultural* (San Francisco: North Point Press, 1981).

Steve Bishop and Christopher Droop, *The Earth is the Lord's: A message of hope for the environment* (Bristol: Regius Press, 1990).

Ian Bradley, *God is Green: Christianity and the environment* (London: Darton, Longman & Todd, 1990).

Susan Bratton, *Christianity, Wilderness and Wildlife: The original desert soltaire* (London: Associated University Presses, 1993).

Seamus Clearly, *Renewing the Earth* (London: CAFOD, 1989).

Tony Compalo, *How to Rescue the Earth Without Worshipping Nature* (Waco, Texas: Word Books, 1992).

Nigel Cooper, *Wildlife in Church and Churchyard* (London: Church House Publishing, 1993).

Tim Cooper, *Beyond Recycling: The longer option* (London: New Economics Foundation, 1994).

Tim Cooper, *Green Christianity* (Sevenoaks, Kent: Hodder & Stoughton, 1990).

Calvin Dewitt, *Earthwise: A biblical response to environmental issues* (Grand Rapids: Christian Reformed Church Publications, 1994).

Calvin Dewitt (ed.), *The Environment and the Christian: What can we learn from the New Testament?* (Grand Rapids: Baker Book House, 1991).

Calvin Dewitt and Ghillean Prance (eds), *Missionary Earthkeeping* (Macon, Georgia: Mercer University Press, 1992).

William Dyrness, *Let the Earth Rejoice: A biblical theology of holistic mission* (Westchester, Illinois: Crossway Books, 1983).

David Ehrenfeld, *The Arrogance of Humanism* (New York: Oxford University Press, 1978).

Dean Freudenberger, *Food for Tomorrow: A Christian agronomist calls for renewal of the biblical covenant to meet the world crisis in agriculture* (Minneapolis: Augsburg Publishing House, 1984).

Dean Freudenberger, *The Gift of Land* (Los Angeles: Franciscan Community, 1981).

Wesley Granberg Michaelson, *A Wordly Spirituality: the call to redeem life on earth* (San Francisco: Harper & Row, 1984).

Wesley Granberg Michaelson (ed.), *Tending the Garden: Essays on the gospel and the earth* (Grand Rapids, Michigan: Erdmans, 1987).

Francesca Greenoak, *Wildlife in the Churchyard: The plants and animals of God's acre* (London: Little & Brown, 1993).

Paulos Mar Gregorios, *The Human Presence: An orthodox view of nature* (Geneva: World Council of Churches, 1978).

John Hart, *The Spirit of the Earth: A theology of the land* (New York: Paulist Press, 1984).

John Houghton, *Global Warming: The complete briefing* (Oxford: Lion Press, 1994).

Mary Jergen and Bruno Manno (eds), *The Earth is the Lord's: Essays on stewardship* (New York: Paulist Press, 1978).

Dale and Sandy Larsen, *Tending Creation* (Wheaton, Illinois: Harold Shaw Publications, 1995).

Sean McDonagh, *The Greening of the Church* (London: Geoffrey Chapman, 1990).

David Mahan, Joseph Sheldon and Raymond Brand, *Redeeming Creation: The biblical basis for environmental stewardship* (Downers Grove, Illinois: Intervarsity Press, 1996).

Jurgen Moltmann, *God in Creation: An ecological doctrine of creation* (London: SCM Press, 1985).

Catherine von Ruhland, *Going Green* (London: Darton, Longman & Todd, 1990).

Colin Russell, *The Earth, Humanity and God* (London: UCL Press, 1994).

Paul Santmire, *The Travail of Nature: The ambiguous promise of Christian theology* (Philadelphia: Fortress Press, 1985).

Francis Schaeffer, *Pollution and the Death of Man: The Christian view of ecology* (Wheaton, Illinois: Tyndale House Publishers, 1973).

John Seymour and Herbert Giradet, *Blueprint for a Green Planet* (Totnes, Devon: Green Books, 1992).

Ron Sider, *Rich Christians in an Age of Hunger* (London: Hodder & Stoughton, 1977).

Tom Sine, *The Mustard Seed Conspiracy* (London: MARC Europe, 1981).

Loren Wilkinson, *Earthkeeping in the Nineties: Stewardship of creation* (Grand Rapids: Eerdtmans, 1991).

Robert Williams (ed.), *Creation, Christians and the Environment: Papers of a Newman Association Conference* (London: Newman Association, 1990).

Nancy Wright and Donald Kill, *Ecological Healing: A Christian vision* (Maryknoll, New York: Orbis Books, 1993).

R. A. Young, *Healing the Earth* (Nashville: Broadman & Holman Publishers, 1994).

The Centre for Alternative Technology

1996 marks the twenty-first anniversary of the Centre for Alternative Technology (CAT). It opened in 1975, well before the ecological message had caught the public eye. CAT set out to develop and prove, by positive living example, new technologies which would provide practical solutions to the problems that are now worrying the world's ecologists.

In the two decades since CAT was founded, there have been rapid developments in the environmental movement as a whole. More and more people from all sectors of society are now saying the essential 'no!' to all kinds of environmental destruction. We have seen success in the worldwide protests against CFCs, nuclear power, unchecked road building, mass whaling and wasting energy.

Recently there has been an even more exciting development in the massive interest in the need to say the essential 'yes!' to alternatives. It is no use simply showing people that the environment is under threat, without showing any practical ways to put things right. CAT's unique seven-acre visitor complex acts as a bridge between those who are seeking to explore more ecological ways of living and the store of practical hands-on experience gained through living and working with sustainable technologies over the past two decades.

On arrival, visitors are carried up a 180ft slope to the visitor complex by a water-balanced cliff railway. The visitor complex is not connected to the National Grid and is powered mainly by a combination of wind, water and solar power. It contains a range of interactive educational displays which are continually updated, reflecting the ways in which society is taking seriously the need to move towards more sustainable lifestyles.

For those who wish to know more, CAT also offers a residential course programme with diverse topics, including small-scale wind and solar power, alternative sewage systems and self-build ecological housing. Tutors have years of practical experience. Their enthusiasm is infectious and courses offer frequent opportunities to talk with them on an informal basis.

To complement the courses, CAT also produces a diverse range of publications covering many aspects of energy, building and food production. These are all written from very practical perspectives, combining a level-headed overview with 'hands-on' information based on CAT's experience of living with such technologies. Each publication also contains an extensive resource guide detailing other organizations working in each area and where to go for more information. Through this practical approach CAT's specialist departments have achieved an international reputation for expertise in sustainable technologies.

Centre for Alternative Technology

Machynlleth, Powys SY20 9AZ, Wales.

Open seven days a week all year round.

Opening hours are 10am to 7pm. Last entry is 5pm.

For further information telephone 01654-702400.

For information on educational or school visits telephone 01654-703743.

Wild Goose Publications would like to thank CAT *for their generous permission to use their illustrations in this publication.*

The Iona Community

The Iona Community is an ecumenical Christian community, founded in 1938 by the late George MacLeod (Very Rev. Lord MacLeod of Fuinary) and committed to seeking new ways of living the gospel in today's world. Gathered around the rebuilding of the ancient monastic buildings of Iona Abbey, but with its original inspiration in the poorest areas of Glasgow during the Depression, the Community has sought ever since the 'rebuilding of the common life', bringing together work and worship, prayer and politics, the sacred and the secular in ways that reflect its strongly incarnational theology.

The Community today is a movement of around 200 Members, over 1400 Associate Members and about 1600 Friends. The Members – men and women from many backgrounds and denominations, mostly in Britain, but some overseas – are committed to a rule of daily prayer and Bible reading, sharing and accounting for their use of time and money, regular meeting and action for justice and peace.

The Iona Community maintains three centres on Iona and Mull: Iona Abbey and the MacLeod Centre on Iona, and Camas Adventure Centre on the Ross of Mull. Its base is Community House, Glasgow, where it also supports work with young people, the Wild Goose Resource and Worship Groups, a bimonthly magazine (*Coracle*) and a publishing house (Wild Goose Publications).

Other titles available from WGP

SONGBOOKS with full music (titles marked * have companion cassettes)

THE COURAGE TO SAY NO; 23 SONGS FOR EASTER AND LENT*
John Bell and Graham Maule

GOD NEVER SLEEPS – PACK OF 12 OCTAVOS* John Bell

COME ALL YOU PEOPLE, Shorter Songs for Worship* John Bell

PSALMS OF PATIENCE, PROTEST AND PRAISE* John Bell

HEAVEN SHALL NOT WAIT (Wild Goose Songs Vol.1)* John Bell and
Graham Maule

ENEMY OF APATHY (Wild Goose Songs Vol.2) John Bell and Graham Maule

LOVE FROM BELOW (Wild Goose Songs Vol.3)* John Bell and Graham
Maule

INNKEEPERS AND LIGHT SLEEPERS* (for Christmas) John Bell

MANY AND GREAT (Songs of the World Church Vol.1)* John Bell (ed./arr.)

SENT BY THE LORD (Songs of the World Church Vol.2)* John Bell (ed./arr.)

FREEDOM IS COMING* Anders Nyberg (ed.)

PRAISING A MYSTERY, Brian Wren

BRING MANY NAMES, Brian Wren

CASSETTES & CDs (titles marked † have companion songbooks)

Tape, THE COURAGE TO SAY NO † Wild Goose Worship Group

Tape, GOD NEVER SLEEPS † John Bell (guest conductor)

Tape, COME ALL YOU PEOPLE † Wild Goose Worship Group

CD, PSALMS OF PATIENCE, PROTEST AND PRAISE † Wild Goose
Worship Group

Tape, PSALMS OF PATIENCE, PROTEST AND PRAISE † Wild Goose
Worship Group

Tape, HEAVEN SHALL NOT WAIT † Wild Goose Worship Group

Tape, LOVE FROM BELOW † Wild Goose Worship Group

Tape, INNKEEPERS AND LIGHT SLEEPERS † (for Christmas) Wild Goose
Worship Group

Tape, MANY & GREAT † Wild Goose Worship Group

Tape, SENT BY THE LORD † Wild Goose Worship Group

Tape, FREEDOM IS COMING † Fjedur

Tape, A TOUCHING PLACE, Wild Goose Worship Group

Tape, CLOTH FOR THE CRADLE, Wild Goose Worship Group

DRAMA BOOKS

EH JESUS...YES PETER No. 1, John Bell and Graham Maule

EH JESUS...YES PETER No. 2, John Bell and Graham Maule

EH JESUS...YES PETER No. 3, John Bell and Graham Maule

PRAYER/WORSHIP BOOKS

PRAYERS AND IDEAS FOR HEALING SERVICES, Ian Cowie

HE WAS IN THE WORLD, Meditations for Public Worship, John Bell

STRANGE FIRE, Ian Fraser

EACH DAY AND EACH NIGHT, Prayers from Iona in the Celtic Tradition, Philip Newell

THE IONA COMMUNITY WORSHIP BOOK

THE WHOLE EARTH SHALL CRY GLORY, George MacLeod

THE PATTERN OF OUR DAYS, Kathy Galloway (ed.)

OTHER BOOKS

COLUMBA: Pilgrim and Penitent, Ian Bradley

THE MYTH OF PROGRESS, Yvonne Burgess

EXILE IN ISRAEL: A Personal Journey with the Palestinians, Runa Mackay

FALLEN TO MEDIOCRITY: CALLED TO EXCELLENCE, Erik Cramb

REINVENTING THEOLOGY AS THE PEOPLE'S WORK, Ian Fraser

PUSHING THE BOAT OUT, New Poetry, Kathy Galloway (ed.)

WHAT IS THE IONA COMMUNITY?

WILD GOOSE ISSUES/REFLECTIONS

WOMEN TOGETHER, Ena Wyatt and Rowsan Malik

THE APOSTLES' CREED: A Month of Meditations, David Levison

SURPLUS BAGGAGE: The Apostles' Creed, Ralph Smith